PHARMA AND PROFITS
BALANCING INNOVATION, MEDICINE, AND DRUG PRICES

John L. LaMattina

Stonington, CT, USA

Registered Office(s)
John Wiley & Sons, Inc., 111 River Street, Hoboken, NJ 07030, USA

Editorial Office
John Wiley & Sons, Inc., 111 River Street, Hoboken, NJ 07030, USA

For details of our global editorial offices, customer services, and more information about Wiley products visit us at www.wiley.com.

Wiley also publishes its books in a variety of electronic formats and by print-on-demand. Some content that appears in standard print versions of this book may not be available in other formats.

Library of Congress Cataloging-in-Publication Data applied for:

Paperback ISBN: 9781119881339

Cover Design: Wiley
Cover Image: © Abscent84/Getty Images

Set in 11/13pt Times LT Std by Straive, Pondicherry, India

SKY10067038_020924

For Sloane, Max, and Coco

CONTENTS

ACKNOWLEDGMENTS

I MUST acknowledge the debt we all owe to the scientists around the globe whose tireless dedication and keen insights have produced the breakthrough vaccines and medications that have led the fight against Covid-19. Countless lives have been saved by their efforts. They deserve our respect and admiration.

I would like to thank a few people for their advice and encouragement during the preparation of this book. Stephen Lederer provided terrific advice on key points in the narrative. The editorial skills of Mary LaMattina improved this manuscript immensely. Finally, Donna Green somehow managed to bring this all together.

J.L.L.

INTRODUCTION

THE WOMAN'S anguished accusation was stunning. "The pharmaceutical industry killed my daughter," she yelled at me. I had just finished taping an episode of the syndicated TV series, *The Dr. Oz Show*, where I had tried, with little success, to defend that same industry . . . once the world's most admired.

In 1997, three pharmaceutical companies were in the top 10 of *Fortune*'s list including Merck (#3), Johnson & Johnson (#4), and Pfizer (#8). Earlier, Merck had been lauded by *Fortune* magazine as the "World's Most Admired Company" for seven straight years.

But by May 2011, when Dr. Oz hosted "The Four Things Drug Companies Don't Want You to Know" that admiration and high regard had vanished. I had been invited to debate Dr. John Abramson, author of *Overdosed America: The Broken Promise of American Medicine*.

Back in the 1990s, those drug companies were best known for their breakthrough medicines for heart disease, depression, AIDS, and bacterial infections. They prospered with products that benefitted hundreds of millions of people. How could they not be admired?

Two decades later, on one of America's most popular TV shows, a fired-up, suspicious audience applauded Dr. Abramson's premise that placed my industry in the same hall of shame as tobacco and oil. The industry had lost its way. And I became even more determined to help overcome this prejudice with counter arguments and data.

On that day, I tried to express my sorrow to the distraught mother, but she abruptly turned and walked away. I learned a few years later from Dr. Abramson that this young girl was being treated for depression and, while on therapy, committed suicide. (I do not know which drug she was taking.) I cannot think of anything more tragic than losing a child. While it is uncertain that the drug was the cause of this suicide, her mother was persuaded. She was unable to accept my sympathy, nor consider my conviction – that the goal of pharmaceutical R&D is to alleviate pain and suffering, not cause it.

My advocacy began in 2006, when I was invited to give a lecture at the University of New Hampshire, which was open not just to the university community but also to the general public. The talk, "Pharmaceutical R&D: The World's Hope for Tomorrow's Cures," was designed to help combat the broad criticism of the industry by answering a number of questions:

- What value do new medicines bring to society?
- Where do medicines come from?
- What innovation does Big Pharma bring?
- How are risks and benefits of medicines evaluated?

This 45-minute talk began at 4 pm with an audience of a few hundred people. The ensuing question and answer period lasted another hour and a quarter and would have gone even longer if the organizers had allowed. The questions covered the entire spectrum of drug R&D.

People were more than curious. They had dozens of questions and wanted answers. They were stunned to hear how long and costly it is to discover and develop new drugs. People had little appreciation of the cutting-edge science needed to be successful. They asked how scientists remain motivated when after spending years on a program, it suddenly dies. By the end of the session, people had a better understanding of the tremendous challenges involved in bringing forward new medicines. The audience began to see a major piece of the healthcare debate in a totally new light.

This experience led me to write *Drug Truths: Dispelling the Myths of Pharma R&D*. As a result of *Drug Truths*, I began to get invitations to speak, not just about pharmaceutical research, but also the industry itself. The call from *The Dr. Oz Show* invited me to debate some of the issues I addressed in my book: the safety of new drugs, the myth that the industry invents diseases, that people are overmedicated, etc. The opportunity to join this discussion appealed to me.

Looking back, I was incredibly naïve. The first time I saw that theme, "The Four Secrets that Drug Companies Don't Want You To Know," was when I walked on stage. Drs. Oz and Abramson were not buying my arguments and nor was the audience.

Confronted with the negative public perceptions of the pharmaceutical industry, I wrote *Devalued and Distrusted: Can the Pharmaceutical Industry Restore Its Broken Image?* It addressed a number of issues that pharma was facing at the time, including: improving its ability to measure not just the benefits but also the risks of new medicines; the need for greater transparency in the conduct as well as the outcomes of clinical trials; changing how drugs are marketed; changing how physicians are compensated for their work with pharmaceutical companies; etc.

Remarkably, the industry has made many changes to improve its accountability – albeit with some prodding from the government. For example, all clinical trials must be registered on the website: *www.clinicaltrials.gov*.

In addition, the results of these studies – positive and negative – must be reported within 12 months of completion. Companies are also making public *any* payment in excess of $10 made to any physician resulting from collaboration with a company. With respect to payments to physicians, the Centers for Medicare and Medicaid Services (CMS) publishes all of these on its *Open Payments* website (*www.CMS.gov/openpayments*).

Despite this progress, there is still one reputational issue that dwarfs all others – the price of drugs. In fact, while our country is divided on almost every problem we face, drug pricing unifies political foes. It even brought together former President Donald Trump and his left-leaning nemesis Senator Bernie Sanders. At a press conference in January 2017, President Trump said: "Our drug industry has been disastrous. . ..And the other thing we have to do is create new bidding procedures for the drug industry, because they are getting away with murder [1]." Senator Sanders responded: "He's right and I've been saying this for years. Pharma does get away with murder. Literally murder. People die because they can't get the prescription drugs that they need [2]."

The pharmaceutical industry should not be surprised by the backlash. After all, stories about high drug prices appear almost daily. How can a single pill cost $1000? How can healthcare systems stay afloat when life-saving gene therapies are priced at over $2 million per patient? Why is Senator Sanders taking busloads of Americans over the border to Canada to get insulin for their diabetes? How can Pfizer and Moderna justify billions of dollars in revenues for their mRNA vaccines? This book seeks to answer these and other questions. Is the drug industry filled with profiteers "getting away with murder" or is it an industry made up of companies that invest in a high-risk business called "innovation" that makes reasonable returns on at-risk capital? You decide.

REFERENCES

1. Karlin-Smith, S. (2017). Trump says drug industry "getting away with murder" Politico, (11 January).
2. Carter, Zachary D. and Schumaker, Erin (2017). Bernie Sanders: Donald Trump is right about big pharma, HuffPost, (11 January).

CHAPTER 1

THE $1000 PILL: THE FISCAL CONSEQUENCES OF CURING HEPATITIS C

"THE COMPANY in this case" is asking for a blank check which if granted will blow up family budgets, will blow up state Medicaid budgets, will blow up employer benefit costs and wreak havoc on the federal debt." This provocative comment was made by Ms. Karen Ignagni, former president and chief executive officer (CEO) of America's Health Insurance Plans, the trade association of health insurance companies. The cause of Ms. Ignagni's alarm was Sovaldi™, a breakthrough drug that cured the liver disease hepatitis C. Manufacturer, Gilead, priced its new medicine at $1000 a pill. Given that the standard course of treatment was once-a-day for 12 weeks, the cost of this cure was $84 000/patient. Ms. Ignagni's concern was shared. Dr. Steven Miller, chief medical officer (CMO) of Express Scripts, a prescription management company, called this drug pricing unsustainable [1].

The World Health Organization (WHO) estimates that globally 71 million people have chronic hepatitis C with roughly 3 million of those in the United States. The most common modes of infection are through exposure to small quantities of blood. While largely asymptomatic, the hepatitis C virus (HCV) resides in the liver and can lead to devastating consequences such as liver scarring, cirrhosis, liver failure, and liver cancer. Many of these patients will ultimately require liver transplants to survive – a surgery that costs more than $300 000.

Older treatments were modestly effective. Cure rates ranged from 40 to 80%, depending on the severity of the disease. Patients were given a cocktail of drugs plus injections of interferon for 24–48 weeks. However, these medicines are poorly tolerated, particularly the interferon component that causes flu-like symptoms in patients. As a result, many with hepatitis C often avoided treatment.

Pharma and Profits: Balancing Innovation, Medicine, and Drug Prices, First Edition. John L. LaMattina.
© 2022 John Wiley & Sons, Inc. Published 2022 by John Wiley & Sons, Inc.

Sovaldi™ provided new hope. The pill was found to cure hepatitis C in more than 90% of patients in just 12 weeks. Furthermore, it is safer and roughly 20% cheaper than the older treatments that cost over $100 000. One would think that the maker of such a wonder drug would be hailed for providing a major medical advance. Instead, Gilead was vilified.

At a *Financial Times* U.S. Healthcare and Life Sciences Conference in New York City, I had a chance to hear Ms. Ignagni talk about the high cost of Sovaldi™. During the Q&A session, I asked her the following question.

> "Sovaldi™ is a drug that cures hepatitis C. It actually *saves* the healthcare system money in that it will prevent patients from dying from liver cancer, cirrhosis and liver failure. Liver transplants alone can cost $300,000 and then patients must take anti-rejection drugs that cost $40,000 per year for the rest of their lives. The price of Sovaldi™, while high now, will drop, first when competitive drugs in late-stage development reach the market and then when the drug is generic. Given all of this, what price for Sovaldi™ would have been acceptable to you – $60,000, $40,000, $10,000? What price are you willing to pay for innovation?"

Ignagni never answered the price question. Instead, she focused on the innovation part, saying that, for years, she has heard that high pricing is needed to sustain innovation. Yet innovation is still occurring. Her response ignores worrying trends that roil the biopharmaceutical industry – the mergers, the small company closings, the reductions in private investment in drug research and development (R&D). Yes, innovation is still occurring, but lower revenues result in less money invested in R&D. Less R&D equals less innovation.

Given the $1000 pill headlines, it was not surprising to see politicians jumping on the bandwagon and expressing outrage over the price. Rather than reacting to this medical breakthrough with applause, this furor sparked Senators Wyden and Grassley to probe all of Gilead's expenses, from the acquisition of Pharmasset (originator of Sovaldi™) to the costs of the development program. Their aim: to embarrass Gilead publicly and, perhaps, shame them into a price cut.

These senators and other politicians have little grasp of the intricacies of drug R&D. Sure, they know R&D is difficult and expensive. They might even appreciate that the entire process, from coming up with the initial idea to getting the US Food and Drug Administration (FDA) approval, can take 15 years. But they have little idea as to how and why drug prices are determined. Do patients or physicians really care how much a company spent in the discovery and development of a new medicine? What they want to know is whether the drug works and, relatively speaking, is it safe? The same can be said of payers. Again, they could not care less about R&D

expenditures. They are much more concerned about the drug's short-term impact on their balance sheet.

Biopharmaceutical companies try to elicit sympathy by talking about failure rates. The industry works on the cutting edge of medical science, looking for novel compounds to prove or disprove medical hypotheses. This is difficult and often frustrating work. Far more projects fail than succeed. Thus, in justifying the high cost of new drugs, companies will cite figures showing that billions of dollars need to be invested across a portfolio of programs to get one new drug approved. Indeed, for a biopharmaceutical company to survive, it has to be profitable. It must provide a return on investment for its shareholders. However, patients, physicians, and payers do not shed tears over a company's litany of failures. The belief is that companies should be rewarded for success, and not for "nice tries."

Thus, in the minds of patients, physicians, and payers, the pricing of drugs should have little to do with the expense of biomedical R&D, nor should it be associated with recouping R&D investment. Pricing should be based on only one thing – the value that the drug brings to healthcare in terms of:

1. Saving lives.
2. Mitigating pain/suffering and improving the quality of life for patients.
3. Reducing overall healthcare costs.

OK, if one were to take that position, how do new, expensive medicines stack up in terms of delivering value? Are they worth the prices sought by drug companies?

In the case of Sovaldi™, we have a drug that cures hepatitis C and, in doing so, prevents the downstream consequences of patients contracting liver cancer or needing a liver transplant. Rather than questioning Gilead's management on the R&D costs generated in Sovaldi's™ development, Senators Wyden and Grassley should ask the following questions:

1. How does this compound compare with existing hepatitis C treatments in terms of efficacy and safety?
2. How expensive are these other treatments?
3. Without this drug, how many patients with hepatitis C will die?
4. How many patients with hepatitis C will contact liver cancer or will need a liver transplant if their disease remains uncured?
5. Without Sovaldi™, what would be the ultimate cost to the healthcare system to treat the resulting cases of liver cancer and liver transplantations?

If these senators do this, they will find that even at $84 000 per patient, Sovaldi™ is well worth the expense. Its value is neither the R&D costs incurred in its discovery and development nor in the years or time and money invested at exploring other approaches to hepatitis C cures that did not work. Its value comes from saving lives and ultimately saving the healthcare system millions. When it comes to drug pricing, to paraphrase a former President Clinton's advisor, James Carville: "It's the *value*, stupid!"

To this point, the focus has been on Sovaldi's™ $84 000 price, which is its *list price*. This number is often the starting point in negotiations with payers to gain access to their formularies and healthcare plans. In the period when only Sovaldi™ was on the market, payers had a relatively weak hand. That, however, all changed when AbbVie launched its hepatitis drug, Viekira Pak™. AbbVie's list price was modestly lower at $83 319 for a course of treatment. However, now payers had a choice and competition can drive costs down. Big payers like Express Scripts and CVS Health do not pay the list price – the one usually quoted by the industry's critics. Instead, given the vast number of patients in their plans, payers can negotiate significant reductions.

At a *Forbes* Healthcare Summit [2] shortly after AbbVie's launch of Viekira Pak™, the aforementioned Dr. Steven Miller, then the CMO of Express Scripts, could not hide his glee that Sovaldi™ now had competition. In fact, he had selected Viekira Pak™ for the 85 million members in the Express Scripts network. At the *Forbes* meeting he surprisingly announced that the price paid in the United States for hepatitis C drugs was less than in Europe. The Express Scripts number remains confidential. However, drug prices in Europe are public. At that time, the United Kingdom paid $55 000/patient for hepatitis C drugs and Germany $67 000. Given Dr. Miller's comments, we can speculate that the US price for these drugs was under $50 000.

After the launches of Sovaldi™ and Viekira Pak™, however, a new generation of hepatitis C drugs became available such as Harvoni™ (Gilead), Zepatier™ (Merck), and Mavyret™ (AbbVie). The last is particularly interesting. AbbVie's list price for the drug, which is approved for all genotypes of hepatitis C, was $26 400 for an eight-week treatment. If the infection was not cured after this time, another four weeks of treatment would cost an additional $13 200 (total of $39 600 for 12 weeks). Suddenly, curing hepatitis C had become relatively cheap.

Against this backdrop, it was surprising to hear Gilead's then CEO, John Milligan, apologize for the original price of Sovaldi™. Milligan's act of contrition was made at another *Forbes* Healthcare Summit [3]. *Forbes* had assembled an impressive panel of biopharmaceutical CEOs to discuss the sorry state of the industry's reputation. Any conversation of this topic immediately turned to drug pricing and Gilead's Sovaldi™ took center stage. Milligan was asked if Gilead's pricing strategy was a mistake.

> "Yeah, it was an interesting launch for the HCV product. We priced the product at exactly the same as the existing standard of care, which worked about 50 percent of the time, and are providing a benefit that, based on real world experience, works about 98 percent of the time. From our perspective, it was a very good value. What happened was a failure to understand exactly how many people were direly ill and had to come into care. That is, there were hundreds of thousands of people who needed this immediately, whose doctors felt that they needed this immediately. The surge into the system was very large, and that created a lot of anxiety around the payers and of course created an outcry against us for having mispriced the product."

As Milligan pointed out, no one questioned the value of the product. Not only are people cured of hepatitis C but an enormous saving is also accrued by the healthcare system down the line as Sovaldi™ reduces the ultimate consequences of HCV: cirrhosis of the liver, liver failure, the need for liver transplants, liver cancer, and, ultimately, death. But despite the tremendous benefits of Sovaldi™, Milligan took a contrite tone:

> "I think our failure, if I have to take a step backwards, was that we were unable to have a good enough conversation with the payers. Perhaps we were a little conservative about what we could have or should have said to them to allow them to prepare for the number of patients that came forward. Honestly, it was far more than we thought. We did not think the system could or would try to handle as many patients as it did. We essentially quadrupled the number of patients treated in a year. That surge really created a lot of pain."

Milligan was being overly apologetic. Payers certainly know how many of the patients in their plans have HCV. Furthermore, the FDA approval of Sovaldi™ was no surprise. The clinical trials for this drug were well publicized, as were the remarkable results. Payers often claim that they will pay for true value. Sovaldi™ certainly demonstrated that both in terms of benefiting patients and offering reduced healthcare costs. This is a life-saving, cost-saving drug. Their outcry against Gilead's list price went against the claims that this is the type of therapy they want the biopharmaceutical industry to produce.

But, again, the discussion focused on Sovaldi's™ *list price*. In fact, as other companies like Merck and AbbVie came up with their own cures, the price to the US payers dropped precipitously. Curing patients for $40 000 or less is a bargain compared to the cost of previous, less effective treatments.

Sovaldi™ is a great example of the value that innovative drugs can bring to us all. Yes, there are examples of terrible gouging, such as irresponsible price hikes for generic drugs or wanton price increases that some companies make for their drugs two or three times a year. But Gilead did not do that. If a company is denied reasonable pricing for excellent new drugs, then the biopharmaceutical industry will likely stagnate and many new opportunities for drug R&D will go unfunded.

This was a great opportunity for Gilead's CEO to voice these issues. It is too bad he did not.

The availability of HCV cures was especially timely for Vietnam War veterans. Many of these heroes contracted the disease as a result of battlefield injuries requiring blood transfusions. They were now suffering from the consequences of this largely silent menace – liver disease, cirrhosis, and liver cancer. Without these drugs, the sicker of these patients were facing certain death.

At the 24th Annual Wharton Health Care Business Conference, Dr. David J. Shulkin, the Secretary of the US Department of Veterans Affairs, announced that the VA was on track to eliminate hepatitis C infections for those willing and able to be treated. In October 2014, the VA had over 146 000 veterans afflicted with hepatitis C. Dr. Shulkin predicted this number would drop to 20 000 [4].

How did this happen? Here is the VA's response as contained in their 2018 Budget in Brief:

> "In 2014, VA began a ground-breaking system of care for Veterans with the Hepatitis C Virus (HCV). The Food and Drug Administration approved two new, highly-effective drugs – Sofosbuvir (Sovaldi™) and Simeprevir (Olysio™) – that work to change the lives of Veterans infected with hepatitis C. Prior to the introduction of the new high-cost treatments therapies in the VA system in January 2014, treatments for hepatitis C were often ineffective and presented considerable side-effects. By contrast, the new treatment options are considerably more effective than earlier options and are much easier to administer. Cure of HCV significantly decreases the risk of progression of the disease to cirrhosis, liver failure, liver cancer, and death. VA wants to ensure that all Veterans eligible for these new drugs, based on their clinician's recommendation, receive the medication" Source: [5] John LaMattina/Forbes Media LLC.

But what about the high cost of these drugs? While the retail price of Sovaldi™ was $84 000 at launch, the VA is allowed by law to negotiate drug prices. In addition, as has been mentioned, other hepatitis C cures have been brought to market over the intervening years such as AbbVie's Viekira Pak™ and Merck's Zepatier™, thus putting purchasers in a good negotiating position. Here is how the VA described drug costs in their 2018 Budget in Brief:

> "VA successfully worked with the manufacturers of these drugs to receive a reduced price for their use to treat Veterans. VA estimates the drugs will cost $748.8 million and provide 31,200 treatments in 2017 and costs increasing to $751.2 million for 28,000 treatments in 2018" Source: John LaMattina [5].

If you quickly do the math, 59 200 US veterans will be cured of hepatitis C for roughly $25 300 for each soldier treated.

This is a great story. Thanks to the VA's commitment as well as the innovation on the part of the manufacturers, a major health issue for our veterans is being eliminated.

While the drop in the cost of hepatitis C drugs has benefited the VA in getting veterans these life-saving medications, not all government agencies can take advantage of this situation. A case in point is the prison system. *Kaiser Health News* [6] reported that 144 000 inmates across 49 states had not been treated with any of the hepatitis C drugs. The reason is cost. Even at $25 000, the price is prohibitive for state prison budgets. The situation in Minnesota is pretty typical. Dr. David Paulson, medical director, was quoted saying: "If we treat everybody with hepatitis C, it would exceed the entire total pharmaceutical budget for everything else and there would not be enough budget left to treat patients with other diseases. We need to do what brings the greatest benefit for the greatest number of people." It is estimated that in Minnesota alone there are 1500 inmates with hepatitis C.

Kaiser estimates that 75 000 hepatitis C patients are released back into the general population annually. As stated before, if not treated, many will progress to cirrhosis, liver cancer, or may even require a liver transplant, all of which would be far more expensive than curing the infection. Furthermore, they could be a source of infection in the community.

Interestingly, two professors, Dr. Anne Spaulding (Emory University) and Dr. Jagpreet Chhatwal (Harvard Medical School), have proposed a solution to this problem – "Nominal Pricing [7]."

> "This pricing mechanism provides deep discounts of at least 90 percent on drugs to so-called safety-net facilities. (By law, the nominal price of a drug must be less than 10 percent of its average market price.) These include hospitals and clinics that treat many patients without insurance or who are homeless. Because of the high markup on hepatitis C drugs, a nominal price is still well above the cost to manufacture the pills.
>
> Nominal pricing permits manufacturers to sell drugs to safety-net facilities at a low price without disrupting the Medicaid market. Offering a discount to one customer usually triggers a similar discount on any drugs sold in the Medicaid program. Nominal pricing provides a way to prevent this from happening, and so can entice drug manufacturers into offering substantially lower prices to safety-net facilities that otherwise may not be able to afford certain drugs" Source: [7]/STAT.

Adoption of such a proposal has obvious benefits. Nominal pricing would bring the cost of treating hepatitis C in prisons to $2500 a patient – much cheaper than treating them after they are released, thus saving the healthcare system a significant sum by avoiding downstream costs. However, putting downstream costs aside, an inmate has a right to medical care. This is the right thing to do.

* * * * * * * * * * * * *

There was a time when curing a disease like hepatitis C would have generated tremendous praise for the pharmaceutical industry. The innovator company would have been lauded for its creativity, insights, and diligence. Interviews would have been conducted with patients who were relieved and grateful that they would not have to endure the consequences of hepatitis C, quite possibly even death. But, the approval of Gilead's Sovaldi™ saw little of that. Instead, everyone focused on "The $1000 Pill."

To a certain extent, that was not a surprise. The industry was getting battered on multiple fronts and this was another opportunity for critics, politicians, and insurance companies to weigh in and focus not on the benefits of the drug, not on the ultimate savings – even at the $84 000 price – that the healthcare system would gain. Rather, the focus was on the drug's list price.

Yet, the cost of Sovaldi™ ended up dropping dramatically. Competition with new hepatitis C drugs enabled payers to negotiate much lower prices, such that the United States was paying less than other nations for these drugs. This is amazing given that countries like the United Kingdom and Germany as single payer nations have remarkable bargaining power since they are buying for their entire populations. Furthermore, at some point, these prices will drop even more dramatically as the patents that currently protect these drugs will expire. Once generic manufacturers enter the picture, curing hepatitis C will likely cost a few thousand dollars a patient.

There are those who will say that, even at a cost of $25 000 per treatment, these drugs are overpriced and that pharmaceutical companies are gouging the public. Yet, there is no evidence of that. If you look at the industry's return on investment or return on capital, this industry's performance is about average across similar sectors. But there is nothing average about the benefits biopharma delivers. As will be seen in subsequent chapters, this is an industry capable of saving the lives of millions of people.

REFERENCES

1. Smith, Michael (2014). AHIP Blasts "Unsustainable" Drug Costs. *Med Page Today*, 21 May.
2. Forbes Healthcare Summit (2015). Better, Cheaper, Safer: Creating the Care We Deserve, 2 December.
3. Forbes Healthcare Summit (2016). Solving Healthcare's Biggest Challenges, 30 November.

4. 24th Annual Wharton Health Care Conference (2018). Adapting to Consumer-Driven Care, 23 February.
5. LaMattina, J. (2018). The VA Will eliminate hepatitis C in veterans by year-end. *Forbes Media LLC.* (1 March 2018).
6. Thanthong-Knight, Siraphob (2018). State Prisons Fail To Offer Cure To 144,000 Inmates With Deadly Hepatitis C. *Kaiser Health News.* 9 July.
7. Spaulding, A.C., Chhatwal, J. (2019). "Nominal Pricing" can help prisons and jails treat hepatitis C without breaking the bank. *STAT.* 9 January.

CHAPTER 2

ENTER THE PAYERS: FDA APPROVAL DOES NOT GUARANTEE COMMERCIAL SUCCESS

THE INTRODUCTION of the statin class of drugs revolutionized the treatment of heart disease. These drugs, notably Lipitor™ and Crestor™, lower the levels of low-density lipoprotein cholesterol (LDL-c), the "bad cholesterol." Numerous studies over the years have shown that these pills reduce heart attacks and strokes in vulnerable populations, that is, those with significant risk factors like atherosclerosis, diabetes, smoking, and a family history of heart disease. In fact, the availability of statins changed medical practice so that physicians now recommend LDL-c levels less than 100 mg/dl [1].

While statins are impressive drugs, a new class of drugs known as PCSK9 inhibitors emerged, which proved to be even more exciting. These are not pills but biological drugs (antibodies) that must be given by injection. They are designed to block an enzyme that mechanistically limits the effectiveness of statins. When PCSK9 antibodies are combined with statins, the LDL-c lowering effects are stunning, with levels as low as 30 mg/dl observed. No other therapy can approach such results. Given the consensus that the lower your LDL-c, the less likely you are to suffer a heart attack or stroke, these results offered great hope to heart patients.

Two PCSK9 antibodies garnered US Food and Drug Administration (FDA) approvals almost simultaneously: Praluent™ (from Regeneron/Sanofi) on 24 July 2015 and Repatha™ (from Amgen) on 27 August 2015. The launches of these drugs should have been celebrated by the cardiovascular medical community. Instead, any enthusiasm was tempered when the prices of these drugs were announced: $14 600 per patient per year for Praluent™ and $14 000 per patient per year for Repatha™. It should be noted that these are list prices. Furthermore, these are not pills but injectable biologics that are expensive to manufacture, store, and distribute. Thus, these prices were expected to be higher than oral statins – but not *that* much higher.

Pharma and Profits: Balancing Innovation, Medicine, and Drug Prices, First Edition. John L. LaMattina.
© 2022 John Wiley & Sons, Inc. Published 2022 by John Wiley & Sons, Inc.

The ensuing outcry was to be expected. Dr. Steve Miller of Express Scripts, echoing his previous views on hepatitis C drugs, said that "Even if physicians adopt this new therapy slower than anticipated, it is clear that PCSK9 inhibitors are on a path to become the costliest therapy class that this country has even seen"[2]. In reality, this did not happen. In fact, 2020 sales of these drugs were quite modest with sales of $887 million for Repatha™ and $358.8 for Praluent™. How could such expensive breakthrough drugs generate such disappointing sales?

Just because a drug has been approved by the FDA does not mean that insurance companies will pay for it. First, there were two PCSK9 inhibitors available, so payers encouraged Amgen and Regeneron/Sanofi to bid against each other for positions on their formularies. Remember, the roughly $14 000 price tag was the *list* price. Details for the deals that companies strike with payers are confidential. Given the choice between Regeneron's Praluent™ and Amgen's Repatha™, Express Scripts opted for the former. While he did not say what Express Scripts was actually paying for Praluent™, Dr. Len Schleifer, chief executive officer (CEO) of Regeneron, glumly told a Forbes Healthcare Summit audience, "Dr. Miller drives a hard bargain." Clearly, Express Scripts as well as other payers were shelling out far less than $14 000 for these medicines.

But beyond negotiating lower prices, payers also limited the ability of patients and physicians to access these drugs. Despite high-risk patients having inadequate LDL-c lowering on statins, studies found 80% of doctors' prescriptions for PCSK9 therapy were denied by payers. After repeated justifications and appeals by the prescribing physician, only 25% of PCSK9 prescriptions were approved by commercial payers and about 50% for Medicare [3]. How could those denials be justified?

There was no doubt that the drugs lowered LDL-c. There were, however, no data that proved that heart attack and strokes were less likely at 30 mg/dL than at 75 mg/dl (which is what is generally achieved with a high dose of a statin like Lipitor™). As a result, payers only allowed these drugs to patients with extraordinarily high LDL-c levels. The concept of "lower is better" for LDL-c to reduce cardiovascular events seemed correct intuitively. However, it was still only a theory.

To prove the true value of the PCSK9 inhibitors in preventing adverse cardiac events, the makers of Praluent™ and Repatha™ had to carry out cardiovascular outcome trials (CVOTs) in patients with known atherosclerotic cardiovascular disease (ASCVD). CVOTs are a huge deal. They generally involve studying anywhere from 10 000 to 25 000 patients over the course of as many as five years. The cost can be $500 million to $1 billion. But such an investment was important to prove the value of PCSK9 inhibitors. A successful outcome in a CVOT would dramatically expand the patient population eligible to receive PCSK9 inhibitors. After all, heart disease is still the leading killer worldwide and these drugs were believed to be major breakthroughs. Thus, cardiologists, heart patients, and payers awaited the results of these CVOTs, albeit for different reasons.

In situations like this, a point often missed is the value that industry-sponsored clinical trials bring to medical science. At this time, no one knew if heart patients would benefit by lowering LDL-c to 30 mg/dl. Only drug companies have the resources to run a CVOT designed to answer such a fundamental biology question. Institutions like the National Institutes of Health do not have the capacity to invest $1 billion for one such trial. Yes, these companies benefit financially with a positive trial, but regardless of the outcome, science benefits, as valuable insights are gained.

The first CVOT to read out was Amgen's study with Repatha™ called FOURIER (Further Cardiovascular Outcomes Research with PCSK9 Inhibition in subjects with Elevated Risk). As often happens in science, the results were not as clear-cut as one would have hoped. Here are some key findings:

- The study enrolled 27 564 men and women, 80% of whom already had a heart attack. The other 20% had experienced a stroke or pain in their limbs due to narrowed arteries.
- Those taking only statins had LDL-c levels on average of 92 mg/dl, which was well within the recommended range. But those on Repatha™ and a statin had an average LDL-c of 30 mg/dl. More impressive was that 25% of those in the Repatha™ arm of the study reached LDL-c levels of 19 mg/dl – an unprecedented number [4].
- Repatha™ reduced heart attacks by 27% (from 4.6/100 patients to 3.4/100), strokes by 21% (from 1.9/100 patients to 1.4/100), and procedures like stents and bypass surgeries by 22% (from 7/100 patients to 5.5).

Those last data concerning cardiovascular events disappointed many experts. They expected a 31% reduction in heart attacks and strokes [5]. Why was there such a difference? Some speculated that Amgen stopped the study at 2.2 years and perhaps, had it gone longer – say five years – a better result would have been observed.

The CVOT carried out by Regeneron and Sanofi for Praluent™, known as ODYSSEY OUTCOMES, had a better result. While there were differences in the study designs (e.g. this trial lasted 2.8 years), the results in some respects were similar to FOURIER.

- There were 18 924 patients enrolled who had experienced an acute coronary syndrome 1–12 months before entering the study and they were equally divided between receiving Praluent™ and placebo along with their maximal statin dose.

- The primary endpoint of this trial was a composite of death from coronary heart disease, nonfatal myocardial infarction, fatal or nonfatal ischemic stroke, or unstable angina requiring hospitalization.
- There were 903 such experiences for those patients on Praluent™ (9.5%) vs. 1052 (11.1%) on placebo. However, Praluent™ showed one big difference. Unlike Repatha™, Praluent™ lowered the overall death rate. While the reason for this difference is not clear, it could be attributed to the longer length of the Praluent™ study [6].

It must be noted that there was another entrant in the PCSK9 antibody race – Pfizer's bococizumab. Pfizer got into the game later than Amgen and Regeneron/Sanofi and, as a result, decided to run its CVOT with bococizumab early in the clinical trial process so its own outcome results would appear at a similar time as the competition.

This was not a trivial decision to make as it called for an upfront investment of hundreds of millions of dollars. However, the early clinical studies looked promising, and this was, after all, an area that was well precedented. What could go wrong?

Stunningly, Pfizer had to stop the trial after 52 weeks [7]. The clinical profile that emerged for bococizumab included an unexpected attenuation of LDL-c lowering over time, as well as a higher level of immunogenicity and a higher rate of injection-site reactions compared with the other PCSK9 inhibitors. None of this was anticipated based on the preclinical and early clinical data already accumulated. As a result, Pfizer dropped bococizumab from development. Overnight, Pfizer lost a promising compound in its pipeline. Hundreds of millions of dollars were sacrificed with no commercial return. Nothing in biomedical R&D is a given.

Still, there were two PCSK9 antibodies that had completed their CVOTs with moderate success. The next obvious question was how best they should be used, especially given their steep price. Typical was the view of Yale cardiologist, Dr. Harlan Krumholz:

> "This study raises the issue of pricing and value. The drug has been priced at about $14,000/year – in part on the promise of a dramatic reduction in risk that was anticipated because of its marked effect on lowering cholesterol. Now that we can see what the drug can achieve, a natural question is: What is this amount of risk reduction, on top of statins, worth?"

At this point, Amgen took an interesting stance on the pricing of Repatha™ in light of the FOURIER results. It offered a "money-back guarantee" for this product [8]. Basically, if a patient has a heart attack while on Repatha™, the payer is eligible for a full rebate from Amgen for the drug's cost.

How does such a program work? Amgen spokesperson Kristen Neese was kind enough to provide me an explanation. In effect, there is not a one-size-fits-all plan. Rather, contracts are negotiated payer by payer. The one requirement is that a patient must be treated with Repatha™ for at least six months before being eligible for full reimbursement. Each contract sets a time limit on how long Repatha™ is expected to protect the heart patient. After all, these are people with heart disease who will likely suffer a fatal cardiovascular event at some point. Repatha™ should not be expected to provide patients immortality. When asked how long the time limit for the money-back guarantee would be, Ms. Neese said that is negotiated with each payer.

She pointed to the Amgen's deal that is in place with Harvard Pilgrim Health Care. In this case, the price that the plan pays is based on how much LDL-c is being achieved with Repatha™. If a patient's LDL-c gets to 30 mg/dl, there is one set price. If the patient's LDL-c fails to get below 60 mg/dl, a lower price is paid. Ms. Neese stressed that the Repatha™ pricing plan is not a gimmick but part of Amgen's commitment to value-based contracting. She believed that, at the time of writing this book, the Amgen plan was unique in the industry.

As governments, payers, patients, physicians, and the biopharmaceutical industry all struggle with pricing of new breakthrough medicines, one would have to think that Amgen's approach will be emulated. If a company wants a high price for a new drug, then it will have to provide some assurances of benefit for patients and value for payers.

Beyond money-back guarantees, competitive factors were driving down the price of PCSK9 inhibitors. The first was a good old-fashioned price war. Amgen began by cutting its price of Repatha™ by 60%, saying that this reduction was geared toward "helping patients afford the medicine at the pharmacy counter." Ms. Neese added that the new $5850 price was in line with the net price that Amgen was receiving after discounts and rebates to pharmacy benefits managers and health insurers [9]. Of course, Regeneron and Sanofi had to respond and, a few months later, the list price of Praluent™ also dropped to $5850. Regeneron CEO, Dr. Len Schleifer, explained the rationale for such a price cut. "In 2018, we lowered the Praluent™ net price for health plans that were willing to improve patient access and affordability. While lowering the net cost to payers did improve access, seniors who were prescribed Praluent™ were often still unable to afford it due to high copay costs or co-insurance at many Medicare Part D plans"[10].

There was, however, more competition on the horizon. The Medicines Company was developing a drug called inclisiran that represented a whole new approach to blocking PCSK9. Licensed from Alnylam, inclisiran (Leqvio™) is an antisense oligonucleotide that targets specific messenger ribonucleic acid (mRNA) sequences thereby disrupting the production of PCSK9 directly in the liver. The first meaningful clinical results for inclisiran were generated in a phase 2 study called ORION-1 and they were impressive [11].

- Total 501 patients at high risk of cardiovascular disease and who had elevated LDL-c were randomized to receive a single dose of placebo or inclisiran (200 mg, 300 mg, or 500 mg) or two doses (at days 1 and 90) of placebo or 100 mg, 200 mg, or 300 mg of inclisiran.
- The primary endpoint of the study was the change from baseline in LDL-c level at 180 days.
- At day 180, mean reductions in LDL-c were 27.9–41.9% after a single dose of inclisiran and 35.5–52.6% for two doses.
- The two-dose 300 mg regimen of inclisiran produced the best results with 48% of patients having LDL-c levels below 50 mg/dl.
- There were no serious adverse events for inclisiran.

Why impressive? After all, the magnitude of LDL-c drops with inclisiran was on the same order of Praluent™ and Repatha™. Inclisiran was, however, going to be a drug that only needed to be dosed twice a *year*, whereas the PCSK9 antibodies need to be dosed twice a month. Given that administration of biological drugs like these can cost over a $1000/visit, the switch to inclisiran could save the healthcare system a lot of money. But the real benefit could be pricing.

An interview with Dr. Clive Meanwell, then CEO of The Medicines Company, on the results of the ODYSSEY OUTCOMES study was quite illuminating [12].

First of all, he believed that Sanofi and Regeneron did a "marvelous trial" that demonstrated the value of PCSK9 inhibitors in treating cardiovascular disease. The data for Praluent™ were presented at the American College of Cardiology (ACC) in 2018. But he was surprised that, on that same day, the Institute for Clinical and Economic Review (ICER), an independent, nonpartisan research organization, published its pricing analysis of these results. (It should be noted that ICER had been highly critical of the initial list prices for Praluent™ and Repatha™.) Here is the ICER summary.

"Based on the results of the ODYSSEY Outcomes trial, ICER has calculated two updated value-based price benchmarks, net of rebates and discounts, for alirocumab (Praluent™) in patients with a recent acute coronary event: $2,300-$3,400 per year if used to treat all patients who meet trial eligibility criteria, and $4,500-$8,000 per year if used to treat higher-risk patients with LDL cholesterol > 100 mg/dL despite intensive statin therapy."

Meanwell was personally pleased that ICER immediately commented on the ODYSSEY results as well as the Praluent™ price reduction. He believed that the launches of Praluent™ and Repatha™ have been "abject failures" and had argued that these drugs need to be priced in such a way as to facilitate availability to patients. It is pretty clear that Sanofi and Regeneron made the ODYSSEY OUTCOMES results available to ICER well ahead of the ACC meeting, in order to help put these data into perspective with respect to value. This proactivity is highly unusual – if not unprecedented. Meanwell said that "ICER has held out an olive branch to change the way drugs are priced" and believed that this type of interaction will be a key component for how drugs will be priced in the future.

However, Sanofi and Regeneron's pricing decision directly impacted The Medicines Company. Meanwell was already planning to broaden access to inclisiran once approved via a significantly lower price than the initial list prices for Praluent™ and Repatha™. He envisioned inclisiran to be the PCSK9 inhibitor for the masses – a drug priced so that anyone with documented ASCVD and anyone who has already had a heart attack would have access to it. Could this propel inclisiran to the front of the line?

Meanwell continued: "Just lowering our price to that of Praluent™ and Repatha™ is not enough. Product differentiation will be important assuming that everyone is operating in the same value window. We will resort to old fashioned product performance. Drugs with the best performance always rise to the top."

His comments were based on the phase 2 results. As we have seen, phase 3 studies can provide surprises – in a negative way. However, that has not been the case for inclisiran's program. Two phase 3 studies particularly stand out – ORION-10 (US based) and ORION-11 (Europe and South Africa based) with the former enrolling 1561 patients with ASCVD and the latter 1627 ASCVD or ASCVD risk-equivalent.

- Patients were randomized to get either inclisiran or placebo on day 1, day 90, day 270, and day 450. The results were as hoped for: ORION-10 showed a 52.3% drop in LDL-c vs. placebo and ORION-11 a 49.9% reduction. Apart from injection-site adverse effects, inclisiran was well tolerated [13].

How important were these results? Well, shortly after these data were made public, Novartis bought The Medicines Company for $9.7 *billion* [14]. Here is what Novartis CEO Vas Narasimhan had to say.

"We are excited about entering an agreement to acquire The Medicines Company as inclisiran is a potentially transformational medicine that reimagines the treatment of atherosclerotic heart disease and familial hypercholesterolemia. With tens of millions of patients at higher risk of cardiovascular events from high LDL-c, we believe that inclisiran could contribute significantly to improved patient outcomes and help healthcare systems address the leading global cause of death. The prospect of bringing inclisiran to patients also fits with our overall strategy to transform Novartis into a focused medicines company and adds an investigational therapy with the potential to be a significant driver of Novartis's growth in the medium to long term."

It is important to note that a CVOT for inclisiran will not be completed until 2024 (ORION-4). The FDA did approve inclisiran (brand name: Leqvio™) at the end of 2021. But, Novartis discarded Meanwell's vision for this drug. CEO Narasimhan believed that Leqvio™ represented a revolutionary approach to treating heart disease and priced it accordingly – $3 250 per dose ($6 500 a year)[15].

Regardless of the Leqvio™ surprise, PCSK9 modulation is a far more economically viable option than the original $14 600 per year price tag. As a result, patients with heart disease are going to benefit. Dr. Steve Miller said that these drugs would become "the most costly therapy our country has ever seen." That has proven to be an exaggeration.

* * * * * * * * * * * * *

There are some similarities between the hepatitis C and the PCSK9 drug stories. Both involved great science leading to true medical breakthroughs that benefit millions of patients around the globe. Yet, the excitement for both breakthroughs was unfortunately muted by price concerns – however justified. In fact, the cost of these drugs was said to threaten the entire healthcare system. But, competition pushed down their prices so it is difficult for insurers to deny access.

But for me, the PCSK9 saga added two new wrinkles. The first was the payers' blatant refusal to allow access to millions of heart patients who could have benefitted from these drugs. Yes, payers had already begun to flex their muscles, for example, in limiting access to specialty rare disease drugs. But this time, cardiologists were greatly dissuaded in prescribing these drugs to those at high risk of a heart attack by being required to submit multiple rounds of paperwork to justify the needs of their patients.

The second stunner was that Amgen began offering money-back guarantees for Repatha™. When you consider that this medicine is prescribed to heart patients who are most likely to eventually die from a cardiovascular event, a money-back guarantee is pretty mind-boggling no matter how the agreement is constructed.

There was a time when a company solely focused on getting an FDA approval for its new drug. Once that approval was in hand, commercial success was a reasonable probability. That is no longer sufficient for a new medical breakthrough. Payers now have a big seat at the table. They can make or break a new product. It is a new world.

REFERENCES

1. LaMattina, J.L. (2009). *Drug Truths: Dispelling the Myths About Pharma R&D*, 16–22. Wiley. Chapter 2.
2. Mangan D.; Tirrell, M. (2015). Pricey new cholesterol Rx covered by big drug plan, but. . .. CNBC.com. 6 October.
3. Kaufman, T.M., Duell, P.B., Purnell, J.Q. et al. (2017). Application of PCSK9 inhibitors in practice. *Circulation Research* 121: 499–501.
4. Kolata, G. (2017).Cholesterol-Slashing Drug Can Protect High-Risk Heart Patients, Study Finds. New York Times. 17 March.
5. Herper, M. (2017). Amgen Drug Prevents Heart Attacks, Not Deaths, Disappointing Experts. Forbes.com. 17 March.
6. Schwartz, G.G., Steg, P.G., Szarek, M. et al. (2018). Alirocumab and cardiovascular outcomes after acute coronary syndrome. *New England Journal of Medicine* 379: 2097–2107. 29 November.
7. Rockoff, J.D. and Hufford, A. (2016). Pfizer cuts outlook, ends development of cholesterol drug. *Wall Street Journal* 1 November.
8. LaMattina, J.L. (2017). Amgen's Money Back Guarantee For Its Cholesterol Drug, Repatha. Forbes.com. 4 April.
9. Beasley, D. (2018). Amgen cuts price of cholesterol drug Repatha by 60%. *Reuters*. 24 October.
10. Herper, M. (2019). Regeneron and Sanofi cut Praluent price 60%, as low-selling drugs become a drug pricing laboratory. *STAT*. 11 February.
11. Ray, K.K., Landmesser, U., Leiter, L.A. et al. (2017). Inclisiran in patients at high cardiovascular risk with elevated LDL cholesterol. *New England Journal of Medicine* 376: 1430–1440. 17 March.
12. LaMattina, J.L. (2018). The Medicine Company CEO, Dr. Clive Meanwell on the Impact of ODYSSEY Outcomes For Heart Patients. Forbes.com. 20 March.
13. Ray, K.K., Wright, R.S., Kallend, D. et al. (2020). Two phase 3 trials of Inclisiran in patients with elevated LDL cholesterol. *New England Journal of Medicine* 382: 1507–1519. 16 April.
14. Novartis Press Release (2019). Company for USD 9.7 bn, adding inclisiran, a potentially transformational investigational cholesterol-lowering therapy to address leading global cause of death, 24 November.
15. LaMattina, J.L. (2022). It Was To Be the Cholesterol Drug for the Masses. Novartis Changed That. 3 January.

CHAPTER 3

PANDEMIC: mRNA VACCINES AND THE RACE FOR A CURE

FOR YEARS scientific experts were worried about a major pandemic engulfing the world. There had already been warning signs. In 2002, an outbreak of severe acute respiratory syndrome (SARS) emerged in China, spread to 8000 people worldwide, and killed nearly 800. Fortunately, for the United States, the disease never gained a foothold in this country. Other pathogens arose such as H1N1 (swine flu), Middle East respiratory syndrome, Ebola virus, and Zika virus. But none of these ever spread globally [1].

In December 2019, there was another pneumonia outbreak in Wuhan, China, but one of mysterious origin. This was a new coronavirus, SARS-CoV-2 (severe acute respiratory syndrome coronavirus 2), and, unlike the previous disease outbreaks, this one spread through the world in a matter of weeks changing our lives dramatically. We went into lockdown. Everything closed. Some businesses never reopened. Grocery deliveries were wiped with disinfectant before being brought into the house. Even the mail was wiped. Where possible, people worked exclusively from home. We were not able to visit loved ones. In the United States, tens of millions would become infected and hundreds of thousands would die. And worst of all, in March of 2020, no one knew how long this situation would last.

President Trump, despite being a known critic of Big Pharma, summoned 10 executives from various pharmaceutical companies to the White House in early March of 2020 to discuss the coronavirus pandemic. He wanted the industry's help in creating a vaccine to prevent the disease as well as therapies to fight it. Trump's behavior mimicked that of most Americans. In a time of medical crisis, we turn to those who can design, develop, and manufacture new medicines and ask: "What can Big Pharma discover to help save lives?"

This event called to mind a similar situation from the 1940s. With the United States in the early stages of World War II, President Franklin D. Roosevelt asked companies to manufacture a miracle drug with the potential

Pharma and Profits: Balancing Innovation, Medicine, and Drug Prices, First Edition. John L. LaMattina.
© 2022 John Wiley & Sons, Inc. Published 2022 by John Wiley & Sons, Inc.

to save the lives of our soldiers on the battlefield. That drug was penicillin, known as a potent antibiotic. But it could only be made in small quantities. It needed to be mass-produced to be truly of value. Nineteen companies heeded Roosevelt's request and ultimately Pfizer, at that time a fermentation company, was able to develop a remarkable process to meet the government's needs. In fact, on D-Day 90% of the penicillin shipped to the war front was manufactured by Pfizer.

Eighty years later, drug companies again mobilized. Within weeks, dozens of companies, from Big Pharma to small biotech, mounted research programs to fight the coronavirus. Researchers sought to find drugs, referred to as therapeutics, to help patients who had already contracted the disease. Gilead was the first with remdesivir, an antiviral drug that reduced the time that Covid-19 patients spent in the hospital. Regeneron also had early success with its Covid-19 antibody cocktail famously used to help treat President Trump [2].

But the Holy Grail was getting a vaccine to people rapidly and there were doubts about how soon one would be available. Ken Frazier, then chief executive officer (CEO) of Merck and highly regarded throughout this field, was worried that a lot of the early hype about vaccines was causing undue optimism. He shared these concerns in an interview with Tsedal Neeley, the Naylor Fitzhugh Professor of Business Administration at Harvard Business School [3].

"There are massive scientific and logistical obstacles to achieving such a feat. What worries me most is that the public is so hungry, is so desperate to go back to normalcy, that they are pushing us to move things faster and faster. Ultimately, if you are going to use a vaccine in billions of people, you'd better know what that vaccine does" Source: [3]/Questex LLC.

He went on to say that it took Merck four years to win approval for its mumps vaccine, at that point the fastest vaccine approval on record. It took five and a half years for Merck to get approval for its Ebola vaccine. Clearly, he believed that a Covid-19 vaccine was not around the corner.

A *New York Times* opinion piece provided data that supported Frazier's view [4]. *The Times* article outlined all of the steps necessary to get a Covid-19 vaccine: preclinical research; phase 1, 2, and 3 trials; building factories; manufacturing; regulatory review; and approval. Even if one compressed these timelines maximally and made investments at risk, such as building factories before you even knew if you had a viable vaccine, the best estimate was that the process would take at least 18 months.

But even before companies had any evidence that they were close to a viable vaccine, cautions were raised about cost. Pharma industry critic

Elisabeth Rosenthal raised a dire specter in a *New York Times* article "How A Covid-19 Vaccine Could Cost Americans Dearly [5]".

> "But a Covid-19 vaccine will have an actual price tag. And given the prevailing business-centric model of American drug pricing, it could well be budget breaking, perhaps making it unavailable to many. . . . . .
>
> Now we are looking for viral deliverance when drug development is one of the world's most lucrative businesses, ownership of drug patents is disputed in endless court battles, and monopoly power often lets manufacturers set any price, no matter how extraordinary. A new cancer treatment can cost a half-million dollars and old staples like insulin have risen manifold in price to thousands of dollars annually.
>
> And the American government has no effective way to fight back.
>
> Recent vaccines targeting more limited populations, such as a meningitis B vaccine for college students and the shingles vaccine for older adults, have a retail cost of $300 to $400 for a full course.
>
> If a Covid-19 vaccine yields a price of, say $500 a course, vaccinating the entire population would bring a company over $150 billion, almost all of it profit."

So, 6 months into the pandemic, people were warning of an 18-month delay before Covid-19 vaccines became available and that their cost will again wreak havoc on the US healthcare system.

And then, amazingly, the big bad biopharmaceutical industry actually saved us.

As companies began their rush to get a Covid-19 vaccine, one would not have necessarily placed a big bet on Pfizer being the first to deliver. Yes, one of Pfizer's biggest products was Prevnar 13™, a valuable vaccine to treat *Streptococcus pneumoniae*. Historically, Merck, GSK, and Sanofi were all bigger players. But Pfizer had a few things going for it. First, it had an outstanding head of vaccine research and development (R&D), Dr. Kathrin Jansen. Second, it already had a research collaboration with BioNTech investigating a promising but unproven technology, messenger ribonucleic acid (mRNA), to create novel vaccines. Finally, Pfizer had a CEO, Dr. Albert Bourla, who was committed to making Pfizer a drug R&D powerhouse.

In February of 2020, when he realized how big a threat this new coronavirus was, Bourla told his leadership team to come up with the best options for Pfizer to pursue and beat the virus. Jansen, working with her BioNTech counterpart, Dr. Ugur Sahin, strongly believed that an mRNA vaccine had terrific potential. In fact, Sahin and his colleagues had already identified potential Covid-19 vaccine candidates [6]. Pfizer and BioNTech announced their collaboration on 17 March.

Actually, Pfizer was not the first company to explore mRNA-based Covid-19 vaccines. That distinction belonged to Moderna, a small biotech company in Cambridge, MA. While in business for over a decade, Moderna had yet to bring a product to market. However, working with the National Institutes of Health (NIH), Moderna was able to design an mRNA vaccine in January of 2020 and began its first clinical trial in March, well ahead of Pfizer's start in May.

For both Pfizer and Moderna, this was going to be a risky endeavor. After all, no mRNA vaccine had been proven to be effective against any virus. Would such a vaccine even work? If so, how effective would it be, keeping in mind that for a vaccine to be of value it has to protect at least 50–60% of those inoculated. Furthermore, creating such a vaccine required huge at-risk investments. First, clinical trials, with tens of thousands of volunteers, were required. Second, the company needed enormous new manufacturing and distribution infrastructure for the doses required by the populations of America and the world. Pfizer estimated its upfront costs at $1 billion. Yes, Pfizer is a big company that had sales in 2019 of $51 billion, but even for a company of Pfizer's size, that was a very expensive bet.

But Bourla did not hesitate.

> "A billion dollars is not going to break us. And, by the way, I don't plan to lose it. I plan to make sure we use this product. You never know until you see the data. So, yes, we are going to lose a billion if the vaccine doesn't work [6]".

Well, that is fine for a maverick CEO to say. But something like this must be agreed by Pfizer's board of directors. True, the pandemic was threatening the health of us all, but this was all unproven and if this investment proved futile it might not "break" Pfizer, but it would certainly damage it significantly. The Pfizer board had to worry about this.

But the board proved to be an asset for Bourla. Of its 16 members, 6 are among the world's leading scientists. Besides Bourla who holds both a DVM and a PhD, the board includes Dr. Susan Desmond-Hellmann (former CEO of the Gates Foundation), Dr. Scott Gottlieb (former commissioner of the US Food and Drug Administration [FDA]), Dr. Helen Hobbs (Professor of Internal Medicine and Molecular Genetics at the University of Texas Southwest Medical Center), Dr. Susan Hockfield (former president of Massachusetts Institute of Technology [MIT]), and Dr. Dan Littman (Professor of Molecular Immunology at New York University [NYU]). These people understood the threat facing us all and the potential for this new technology. They provided strong support for Bourla's plan as Gottlieb shared with Forbes:

> "Albert laid out early why it was so important to put up the enormous resources of Pfizer without an eye toward the business bottom line. Coming up with a vaccine could change the course of human history. That's literally what's at

stake, and big companies have the ability to scale up manufacturing and run big trials in a way not available to small product developers."

It is a good thing that Bourla had the board's backing. When the dust eventually settled, Pfizer invested close to $2 billion in its Covid-19 vaccine program. That, too, would not have broken Pfizer. But shareholders would have taken their revenge.

As companies ramped up their efforts to deliver a vaccine, Operation Warp Speed (OWS), a partnership of the Department of Health and Human Services (HHS), the Department of Defense (DOD), and the private sector, set out to pave the way for the rapid development of multiple vaccines. Dr. Moncef Slaoui, the OWS head, laid out the goals [7].

"The initiative set an ambitious objective: to deliver tens of millions of doses of a SARS-CoV-2 vaccine – with demonstrated safety and efficacy, and approved or authorized by the FDA for use in the U.S. population – beginning at the end of 2020 and to have as many as 300 million doses of such vaccines available and deployed by mid-2021. The pace and scope of such a vaccine effort are unprecedented."

OWS selected vaccine candidates based on four criteria: robust preclinical activity; potential to enter large phase 3 efficacy trials in the July to November timeframe, vaccine technology that allowed for rapid and efficient manufacturing, and utilization of vaccine technologies believed to be most likely to yield a safe and effective vaccine. Ultimately, OWS executed partnerships with six companies: Moderna and Pfizer/BioNTech, which employed mRNA technology; AstraZeneca and Johnson & Johnson (J&J), which utilized a replication-defective live vector platform; and Novavax and Sanofi/GSK, with a recombinant-subunit-adjuvanted protein technology.

In the early summer of 2020, OWS began to sign contracts with these companies for Covid-19 vaccines. Again, these contracts were predicated on demonstrating that a company's vaccine was both safe and effective and that it received an Emergency Use Authorization (EUA) from the FDA. The biggest deal crafted by OWS in terms of numbers of vaccines was with AstraZeneca, which agreed to provide 300 million doses for $1.2 billion dollars.

But, the US government also signed contracts for 100 million doses each from J&J ($1 billion), Novavax ($1.6 billion), the joint effort by Sanofi and GSK ($2.1 billion), Moderna ($1.5 billion), and Pfizer/BioNTech ($1.9 billion). In addition, the United States also provided significant funds to help in the development of these vaccines, particularly for building manufacturing capabilities. For example, Moderna received an infusion of almost $1 billion and J&J $450 million for these purposes.

Absent from the list of companies getting money from OWS for vaccine R&D was Pfizer. (There were some industry critics who mistakenly claimed that Pfizer received $1.9 billion from OWS to support vaccine development. However, those funds were only to be paid upon delivery of 100 million doses of a Covid-19 vaccine after EUA from FDA.) Dr. Mikael Dolsten, Pfizer's Chief Scientific Officer and President of Worldwide R&D, explained why Pfizer abstained [8].

> "This is the worst pandemic in 100 years; Pfizer should be one of the companies that can stand up and help stop it. It is not about seeking funding from the government. It is our job to put our capital at risk if we think we can do something extraordinary. We funded $2 billion at risk (to work with BioNTech and develop/produce their never before used mRNA Covid-19 vaccine technology), but it allowed us to move faster, be more nimble, be less dependent on bureaucracy and a huge chain of structures."

With OWS support, companies were working 24/7 to deliver. However, this was an election year in the United States and politics began to intervene. The United States was in the midst of a contentious election and the Trump administration was eager to have the vaccine delivered before Election Day to help ensure victory. Feeling undue political pressure, nine drug companies took the extraordinary step of issuing a pledge that they would not release any vaccine before it had been fully vetted by the FDA. In a press release entitled "Biopharma Leaders Unite To Stand With Science," the companies made the following pledge [9].

> "We, the undersigned biopharmaceutical companies, want to make clear our on-going commitment to developing and testing potential vaccines for Covid-19 in accordance with high ethical standards and sound scientific principles.
>
> The safety and efficacy of vaccines, including any potential vaccine for Covid-19, is reviewed and determined by expert regulatory agencies around the world, such as the United States Food and Drug Administration (FDA). FDA has established clear guidance for the development of Covid-19 vaccines and clear criteria for their potential authorization or approval in the U.S. FDA's guidance and criteria are based on the scientific and medical principles necessary to clearly demonstrate the safety and efficacy of potential Covid-19 vaccines. More specifically, the agency requires that scientific evidence for regulatory approval must come from large, high quality clinical trials that are randomized and observer-blinded, with an expectation of appropriately designed studies with significant number of participants across diverse populations" Source: [9]/Pfizer Inc.

The signatories were AstraZeneca, BioNTech, GSK, J&J, Merck, Moderna, Novavax, Pfizer, and Sanofi. These companies collectedly had already developed more than 70 novel vaccines. They hoped that, despite the political climate, their pledge would boost public confidence that Covid-19 vaccines had known safety and efficacy. The pledge had its desired effect as evidenced by a perspective published in the *New England Journal of Medicine* by two prominent industry critics: Dr. Jerry Avorn and Dr. Aaron S. Kesselheim. In their aptly titled "Up Is Down – Pharmaceutical Industry Caution vs. Federal Acceleration of Covid-19 Vaccine Approval," they shared the following views [10].

> "For the companies as for the public, a careful data-driven decision in 2021 would be far better than a few months' more sales after a premature approval in October.
>
> FDA guidance issued on June 30 stated that for approval, a vaccine should demonstrate a 50 percent or greater reduction in the incidence or severity of Covid-19 infection as compared with placebo, a relatively high bar. The necessary combination of person-time and disease incidence is daunting. . .
>
> Despite their pledge's potential vulnerability to regulatory indiscretion by political leadership, the pharmaceutical executives are to be commended for the science-based position – which is in their best interests as well. The federal response will help determine not just the fate of the first Covid-19 vaccines, but also what remains of the public's eroding trust in one of the most important aspects of U.S. science policy."

Other opinion leaders shared these views. The pledge helped restore some confidence in the vaccine approval process. At the time it was issued, three companies had Covid-19 vaccines in late-stage clinical trials: Pfizer/ BioNTech, Moderna, and AstraZeneca. Now, the world waited for these clinical trial results.

On the morning of Monday, 9 November 2020, Americans woke up to their first good news in eight months. The Pfizer/BioNTech Covid-19 vaccine worked – and worked spectacularly. The FDA's original bar for approval was set at 50%, but the unprecedented mRNA vaccine was at least 90% effective [11].

Pfizer had only learned of the results on the day before. The independent Data Monitoring Committee (DMC) did their first interim analysis of Pfizer's phase 3 clinical study and the DMC shared the results with Pfizer leadership via a Zoom call. Pfizer had given *National Geographic* exclusive access to the entire vaccine R&D process and so this Zoom call has been captured for posterity in a documentary called "Mission Possible: The Race for a Vaccine." The tension in the room was palpable as the executives nervously awaited the DMC's findings. With the words: "Good news, we made it!" the room erupted. When told that the vaccine was 90% effective, Pfizer CEO Bourla exclaimed: "We have a f***ing vaccine!" (This is my favorite biopharma CEO quote of all time.)

No one expected such efficacy, especially from a totally new type of vaccine utilizing mRNA. Dr. Anthony Fauci called the results "just extraordinary," adding that "Not very many people expected it would be as high as that." For comparative purposes, the annual flu vaccine is usually only about 60% effective. This efficacy was not unique to Pfizer's vaccine. A week later Moderna and the National Institutes of Health (NIH) reported that its Covid-19 vaccine also showed efficacy of almost 95% [12]. These results were from large studies with more than 43 000 in the Pfizer trial and more than 30 000 in Moderna's – a remarkable recruiting achievement accomplished in a matter of months.

All during these clinical trials, and those of others developing Covid-19 vaccines, the FDA worked closely with each drug sponsor in order to optimize the review process as these trials progressed. Within a month of Pfizer's announcement of the results from its interim analysis, the FDA called a meeting of the Vaccine and Related Biological Products Advisory Committee (VRBPAC), which was charged with deciding whether the benefits of the vaccine outweighed its risks. Just as the companies were driving hard to bring forward a vaccine, the FDA was also working 24/7 doing its part to fight the pandemic in a remarkable display of cooperation. On 10 December, the VRBPAC did indeed recommend that the FDA grant an EUA for the Pfizer/BioNTech vaccine [13]. On 18 December, the VRBPAC also recommended the Moderna vaccine for an EUA. On 14 December, healthcare workers were the first to receive their first dose of the Pfizer vaccine and a week later Moderna followed. A mere nine months after the country was shut down because of the pandemic, the biopharmaceutical industry delivered two totally novel vaccines thus bringing hope to the world (Table 3.1). Science did win.

TABLE 3.1   mRNA timeline for Moderna and Pfizer

10 January	SARS-CoV-2 virus sequenced
15 January	NIH designs mRNA vaccine in collaboration with Moderna
16 March	Moderna phase 1/2 trial begins
2 May	Pfizer/BioNTech phase 1/2 trial begins
14 July	Moderna phase 1/2 trial published in *New England Journal of Medicine*
27, 28 July	Moderna and Pfizer/BioNTech phase 3 trial begins
12 August	Pfizer/BioNTech phase 1/2 published in *Nature*
22, 27 October	Enrollment in both phase 3 trials complete; >74 000 participants
9 November	Pfizer/BioNTech announces interim analysis efficacy >90%
16 November	Moderna announces interim analysis efficacy 94.5%
18 November	Pfizer/BioNTech announces 95% efficacy at final result
20 November	First EUA submitted by Pfizer/BioNTech
27 November	Distribution of vaccine by UAL charter flights throughout the United States
10 December	FDA external review of Pfizer/BioNTech EUA
11 December	Phase 1a vaccination begins for healthcare professionals

J&J was the next company to receive an EUA for its Covid-19 vaccine, getting approval on 27 February 2021. J&J chose to develop its vaccine as a single shot. In addition, this vaccine did not require the cold storage temperatures needed for the mRNA vaccines. These features made J&J's vaccine attractive for use in regions where vaccine distribution would be a problem. But J&J had two issues. The first was the vaccine's efficacy, which J&J's clinical trial data showed to be 72%. Had J&J been the first Covid-19 vaccine to generate results, the 72% would have been hailed as a great result. However, in the face of the outstanding potency of the mRNA vaccines, J&J's was perceived by some as an inferior option. The second issue was manufacturing. Unlike Pfizer and Moderna, J&J took a conservative approach to manufacturing and did not begin scaling up to meet the OWS goal of 100 million doses until the back end of its contract. By the time J&J delivered on the bulk of its obligation, the United States already had enough mRNA vaccines to inoculate 200 million Americans. The J&J vaccine proved to be a valued addition to the US pandemic arsenal, but its use was dwarfed by the first entrants [14].

The more curious case was that of the AstraZeneca Covid-19 vaccine.

When OWS was planning for the eventual availability of Covid-19 vaccines, the biggest order by far was for 300 million doses for the AstraZeneca vaccine. This disproportionate order was justified by the belief that this vaccine would be much cheaper than the others and that it would possibly be able to be stored in refrigerators and not the extremely low temperatures that would be required for the Pfizer/BioNTech and Moderna vaccines.

Unfortunately, AstraZeneca was hit with troubles in its large-scale phase 3 clinical trials. First, its clinical program was put on hold for seven weeks while AstraZeneca clinicians investigated the occurrence of what was described as a neurological condition, consistent with transverse myelitis, in two study participants. After closely looking into the data, keeping in mind that transverse myelitis can be caused by a variety of factors not related to the vaccine, the FDA allowed AstraZeneca to continue its program.

In late November 2020, AstraZeneca released the results of its study that proved confounding. Patients who received two full doses of the AstraZeneca vaccine showed efficacy of 62%. However, for those patients who mistakenly received a half-dose followed by a full dose, efficacy was 90%. Had this error not been made, reviewers would have come away with the impression that this vaccine was quite inferior to those of Pfizer/BioNTech and Moderna, which yielded efficacies of about 95%. But, there were other concerns raised by this study. First, only about 2800 patients had received the low-dose, high-dose regimen – not enough to make a conclusive assessment. More troubling was that no one in that arm of the study was over 55. Clearly, AstraZeneca had to run another study to convince the FDA that it truly had a viable dosing protocol particularly for people over 55.

In the meantime, Pfizer/BioNTech, Moderna, and J&J launched their vaccines and more than half of the population had been vaccinated in the United States. AstraZeneca was yet to complete its filing with the FDA. While on hold in the United States, AstraZeneca launched its vaccine in December 2020 in the United Kingdom, where a proud nation purchased 100 million doses of its homegrown vaccine. The AstraZeneca vaccine was also made available in Europe where, unfortunately, new safety concerns arose with reports that this vaccine was associated with severe blood clots. This resulted in the temporary suspension of its use in Denmark, Norway, and Iceland.

Given all that went on, it was no surprise that people shied away from the AstraZeneca vaccine. In Germany, a country anxious for vaccinations, people exhibited a strong preference for the Pfizer/BioNTech vaccine. As reported by Melissa Eddy in the *New York Times*, two weeks after the delivery of 1.45 million doses of the AstraZeneca vaccine in late February, only 270 986 had been administered. A survey showed that, while 81% of Britons believed that the AstraZeneca vaccine is safe, only 43% of Germans and 33% of the French considered it so [15].

Turning back to the United States, what is the Biden administration going to do about the AstraZeneca Covid-19 vaccine? Chances are that, even if emergency use was granted by the FDA, this vaccine will be viewed as inferior by Americans. Given the bounty of vaccines in the United States, Americans will not be clamoring for AstraZeneca's. With the success of this vaccine in the United Kingdom and the need for vaccines in other parts of the globe, distributing the AstraZeneca vaccine elsewhere makes a lot of sense.

While AstraZeneca had a torturous path to getting its vaccine, it did eventually succeed. However, three of the biggest vaccine makers, Merck, GSK, and Sanofi, did not. Looking back, did OWS err in placing its bets? No, but there are some lessons for us from its initial decision.

First, OWS took a "shot on goal" approach with respect to Covid-19 vaccines. That proved to be important as not all companies were successful. However, OWS should not have relied so heavily on the AstraZeneca vaccine at first but rather placed equal orders to maximize the opportunity for success. Fortunately, Pfizer/BioNTech were able to scale up production of their vaccine quickly to make up for the shortfall, otherwise the United States would not be in the enviable position it is today.

Second, despite all of the experience that the world has in developing vaccines, this type of R&D is still very hard – as evidenced by the lack of success from Merck, Sanofi, and GSK, as well as the production challenges experienced by the others. Clearly, it was important for the world to have many companies working on finding a Covid-19 vaccine as opposed to having one global concerted effort. Some lobbied for the latter and that would have been a disaster.

Finally, government funding is critical in a pandemic. While Pfizer/BioNTech did not take US funding, the others did. There was no way Moderna could have succeeded without such funding, nor would the other companies have been able to move forward. There are some who have challenged the use of government funds to support a private enterprise but, in a pandemic, is there any better use of government funding?

"Some Vaccine Makers Say They Plan to Profit From Coronavirus Vaccine" read the *New York Times* headline back in July 2020 [16]. The thought that companies would spend billions of dollars while shifting the focus of their R&D organizations to discovering and developing a Covid-19 vaccine and *not* make a profit would seem odd to many, particularly in the business world. Yet, two companies, J&J and AstraZeneca, said that if successful they would sell their vaccine at cost. Pfizer and Moderna made no such pledge.

Actually, the price of these vaccines was set with the contracts that each company signed with OWS. Pfizer was the most expensive with $1.95 billion for 100 million doses ($39 for the necessary two-dose regimen per person). AstraZeneca was the cheapest at $1.2 billion for 300 million doses ($8 per two-dose regimen). Pfizer CEO Bourla explained the rationale that went into Pfizer's vaccine price [17].

> "Back in June of 2020 we decided to offer our vaccine through tiered pricing. The wealthier nations would have to pay in the range of about the cost of a takeaway meal and would offer it to their citizens for free. The middle-income countries were offered doses at roughly half that price and the low-income countries were offered doses at cost. Many of the poorest communities will receive their doses through donation. Equity does not mean we give everyone the same. Equity means we give more to those that need more."

Bourla went on to say that Pfizer expected to produce 3 billion doses of vaccine in 2021 and another 4 billion doses in 2022; 40% of the 2021 doses were to go to middle- and low-income countries. Given that the world's population is estimated to be about 7.9 billion people, Pfizer by itself will produce enough vaccines to inoculate 45% of the globe – a stunning achievement.

To put the Pfizer vaccine price into perspective, the price of your annual quadrivalent flu shot obtained at a local pharmacy can range from $35 to $41. The more potent flu shot mandated for people 60 and older costs *double* that. It has already been mentioned that vaccines for meningitis B and shingles cost in the hundreds of dollars for a full course of therapy. Quelling a pandemic at $39/person is a bargain.

There is a concern as to how long Pfizer will maintain the $39 price. These concerns were ignited by remarks made by Pfizer chief financial officer (CFO) Frank D'Amelio at Barclays Global Healthcare Conference [18].

> "In terms of pricing let me see if I can hit on that. So, if you look at how current demand and current pricing is being driven, it's clearly not being driven by what I'll call normal market conditions, normal market forces. It's really been driven by kind of the pandemic state that we've been in and the needs of government to really secure doses from the various vaccine suppliers.
>
> So, what we believe, what I believe is as we move from a pandemic state, from a pandemic situation, to an endemic situation normal market forces - normal market conditions - will start to kick in. And factors like efficacy, booster ability, clinical utility will basically become very important and we view that as quite frankly a significant opportunity for our vaccine from a demand perspective, from a pricing perspective given the clinical profile of our vaccine. So clearly more to come here. But we think as this shifts from pandemic to endemic we think there's an opportunity here for us" Source: [18]/Forbes Media LLC.

Post-pandemic will there be a price surge for Covid-19 vaccines? If people will need annual Covid-19 vaccinations as we have for influenza, will the prices double or triple? D'Amelio believes that prices will increase and that such increases will be based on "factors like efficacy, booster ability and clinical utility." There is some truth to that. But first we need data that compare not only the vaccines we now have but also other vaccines that might still be authorized, such as that from Novavax.

So, as we shift to an endemic Covid-19 world and we have the choice of multiple Covid-19 vaccines, how will the government decide which ones to acquire? The driver is likely to be cost. J&J's vaccine is already one-third the cost of Pfizer's and Moderna's and, without head-to-head comparisons, it is hard to say that its 72% efficacy rates are meaningfully lower than its predecessors. Competition is becoming more and more important when payers are considering which drugs to put on their formularies. This will also be true for Covid-19 vaccines even when we are in an endemic world. It is hard to envision huge price increases for Covid-19 prices in the future.

Interestingly, biopharmaceutical companies are not the only ones profiting from Covid-19 vaccines.

On Valentine's Day, I was fortunate to get my first Covid-19 vaccine shot – Pfizer, of course. I was vaccinated at the Floyd Little Fieldhouse in New Haven, which was set up by Yale New Haven Health. It was an excellent experience. The staff and volunteers were polite, helpful, and well organized. The entire process took less than 30 minutes and that included the obligatory 15-minute wait after the vaccination to be sure that there were no immediate adverse effects. When I went back three weeks later, the process was just as efficient. Given a capacity of 1400 people a day, Yale New Haven Health did a terrific job.

These vaccines are free to all patients, thanks to funding by the US government that negotiated contracts with the Covid-19 vaccine makers that will supply enough vaccines to cover our entire population. Other costs related to the Covid-19 vaccination process had been a bit of a mystery to me, until I received my quarterly Medicare Part B summary, which contained the following information. (Keep in mind that I was not billed for these charges.) Medicare reimbursed Yale New Haven Health $18.79 for my first shot and $30.92 for my second. While Pfizer, the innovator and risk taker, received $39 for my vaccination, Yale received $49.71. Multiply this by the thousands that went through the Floyd Little Fieldhouse, it is clear that Yale New Haven Health did pretty well financially.

This is not unique for hospital systems. According to the Centers for Medicare and Medicaid Services (www.CMS.gov), as of 12 August 2021, the Medicare payment rates for administering the vaccines are approximately $40 for single-dose vaccines and, for vaccines requiring multiple doses, approximately $40 for each dose in the series, including any additional doses. Consistent with this, my booster shot which I received at a local Walgreens cost Medicare $43.

For those who are administering Covid-19 shots to be profiting more than the companies that actually brought us these wonderful vaccines is mind-boggling.

As Pfizer progressed in its work to develop a vaccine, in June of 2020 it began to approach countries to provide them the opportunity to pre-order vaccines. Of course, this was dependent on Pfizer's efforts being successful. Recognizing that it would not be able to fulfill the world's vaccine needs in the first months of production, Pfizer wanted to be prepared to honor requests for shipments based on those who expressed early interest. Obviously, OWS ordered 100 million doses but other countries also signed contracts including the United Kingdom, Japan, and Canada.

One country that deferred was Australia [19]. In the summer of 2020, the pandemic was largely controlled in Australia and the country was not in dire need for a vaccine, particularly one that was projected to be as expensive as Pfizer's. Instead, Australia decided to bank on the AstraZeneca vaccine as well as a vaccine that was being developed at the University of Queensland. Both offered the advantage of being able to be produced locally at a facility in Melbourne.

Unfortunately, the University of Queensland's effort proved unsuccessful and, as mentioned earlier, AstraZeneca's was hit by delays. As a result, Australia was left in the lurch. In November, when the outstanding results for Pfizer's vaccine were announced, Australia finally put an order in for the Pfizer vaccine – but only for 10 million doses. Given that the population of Australia is 25 million and that two doses of the mRNA vaccine are needed, only 20% of Australians would be getting access to the Pfizer vaccine. Meanwhile, the demand for the Pfizer vaccine soared. Australia did order another 20 million doses in February but, given the existing orders from other countries, Australia's request was far back in the queue.

As of September 2021, Australia was in lockdown and that plan was to remain so until 70% of its population was fully vaccinated [20]. The government hopes that with the doses of both the Pfizer and finally the AstraZeneca vaccines, the country will be able to meet its inoculation goals. It also hopes to have the Novavax vaccine available to Australians once it is approved.

Australia had an advantage over most of the world as the pandemic appeared relatively later there. Had it been aggressive with ordering vaccines as had happened in the United States with OWS, it probably would have been able to avoid locking down the country. Yes, it might have proved costly. However, ordering 40 million doses of the Pfizer vaccine would have been enough to vaccinate 80% of Australians. That would have cost US $780 million. Looking back on Australia's plight and the economic costs of the lockdown, that would have been a worthwhile investment.

* * * * * * * * * * * * *

The Axios-Harris Poll 100 annually produces a ranking of the most visible US companies with respect to reputation. The survey's Reputation Quotient ranking is based on company performance in seven key areas: trust, vision, growth, products and services, culture, ethics, and citizenship [21]. Published in May 2021, there were two notable entries in the top 10: Moderna, which was new to the list, at number 3 and Pfizer at number 7, which had the largest year-over-year gain in reputation of any company. Clearly, the public recognized the historic achievement of these companies that delivered Covid-19 vaccines in nine months.

While Moderna and Pfizer led the way in this survey, other surveys confirmed that the biopharmaceutical industry in general was now far better regarded by the public than the depths it had fallen pre-pandemic. Dozens of companies looked not just for vaccines but also therapeutants to treat those who became ill with Covid-19 and these efforts were applauded by the public. Drug companies were no longer reputationally lumped with tobacco and oil. While it was uncertain that the world would be completely rid of SARS-CoV-2, there was a sense that we would be able to adapt to the presence of this coronavirus perhaps with the need for annual inoculations much as happens with the influenza virus. The dire state that we were in in March of 2020 has thankfully passed. The biopharmaceutical industry made this happen.

However, the biopharmaceutical industry was not universally praised. The familiar bugaboo – price and profits – again reared its head. Are such criticisms justified? That will be addressed next.

REFERENCES

1. Szabo, L. (2020). Many US health experts underestimated the Coronavirus. . .Until it was too late. *KHN* (21 December).
2. Cohen, J. (2020). Update: here's what is known about Trump's Covid-19 treatment. *Science* (5 October).
3. Sagonowsky, E. (2020). Merck CEO Frazier says Covid-19 vaccine hype a 'grave disservice' to the public. *Fierce Pharma* (14 July).
4. Thompson, S.A. (2020). How long will a vaccine really take? *New York Times* (30 April).
5. Rosenthal, E. (2020). How a Covid-19 vaccine could cost Americans Dearly. *New York Times* (6 July).
6. Vardi, N. (2020). The man betting $1 billion that Pfizer can deliver a vaccine by this fall. *Forbes*, (20 July).
7. Slaoui, M. and Hepburn, M. (2020). Developing safe and effective Covid vaccines – Operation warp speed's strategy and approach. *New England Journal of Medicine* 383 (18): 1701–1703. 29 October.
8. Adams, B. (2021) Pfizer's R&D President Dolsten: not taking government cash made us more nimble for Covid-19 vaccines. *Fierce Biotech* (22 January).
9. Pfizer Press Release (2020). Biopharma leaders unite to stand with science. (8 September).
10. Avorn, J. and Kesselheim, A.S. (2020). Up is down – Pharmaceutical industry caution vs. federal acceleration of Covid-19 vaccine approval. *New England Journal of Medicine* 338 (18): 1705–1708. 29 October.
11. Pfizer Press Release (2020). Pfizer and BioNTech announce vaccine candidate against Covid-19 achieved success in first interim analysis from phase 3 study, (9 November).
12. McNamara, D. (2020). Moderna Covid-19 vaccine: interim data show 94.5% efficacy. *MedScape.com* (16 November).
13. Inserro, A. (2020). FDA advisory panel recommends Pfizer/BioNTech Covid-19 vaccine. *AJMC* (10 December).
14. Weiland, N.; LaFraniere, S. (2021). FDA clears Johnson & Johnson's shot, the third vaccine for U.S. *New York Times* (27 February).
15. Eddy, M. (2021). Germans Clamor for Covid vaccines, but Shun AstraZeneca's offering. *New York Times* (25 February).
16. Wu, Katherine J. (2020). Some vaccine makers say they plan to profit from coronavirus vaccine. *New York Times* (21 July).
17. Pfizer Press Release (2021). An Open Letter from Albert Bourla to Pfizer Employees. (21 March).
18. LaMattina, J.L. (2021). How will Covid-19 vaccines be priced post-pandemic? Forbes. com (23 March).
19. Knaus, C.; Evershed, N. (2021). "Stuffed": how Australia's unconscionable gamble on Covid vaccines backfired. *The Guardian* (9 July).
20. Kelly, L. (2021). Australian Prime Minister Morrison defends lockdown strategy as daily cases hit record. *Reuters* (22 August).
21. The Harris Poll (2021). Patagonia Honda, Moderna, Chick-fil-A, SpaceX Top Axios-Harris Poll 100 With the Best Reputations. (13 May).

CHAPTER 4

FEDERAL INVESTMENT IN R&D: WHY THE GOVERNMENT DOES NOT DESERVE A PIECE OF BIOPHARMA'S PROFITS?

O<small>N</small> **28** J<small>ULY</small> 2021, Pfizer projected sales of its Covid-19 vaccine, Comirnaty™, would be roughly $36 billion for the year. (Pfizer shares its Covid-19 vaccine revenues with BioNTech [1].) To put this daunting number into perspective, Pfizer's total sales in 2019 were $51.7 billion. Its biggest selling product prior to Comirnaty™ was Lipitor™, which had peak sales of $12.9 billion in 2007. Comirnaty™ alone would rank in the top 10 of all pharmaceutical companies in terms of revenues. And these sales are being generated despite the fact that this vaccine is being sold at cost in parts of the world. Furthermore, after manufacturing 3 billion doses of Comirnaty™ in 2021, Pfizer anticipates producing another 4 billion in 2022. Clearly, 2021 was not a one-time phenomenon.

Shortly after these numbers were reported, I was asked by an interviewer: "How much of this money should Pfizer be allowed to keep?" Huh?

Actually, challenging the merits of biopharmaceutical company profits is nothing new. The argument often raised is this industry's success relies largely on the largesse of the National Institutes of Health (NIH) and the academic research it supports. The implication, of course, is that Americans are not getting a fair deal when it comes to drug pricing as their tax money is being used to fund research relied upon by the industry. Dr. Marcia Angell (Harvard Medical School) was particularly vitriolic on this point where her lack of respect for industry was quite apparent in a *Washington Post* op-ed [2].

> "[Drugs] are invented not by the companies that sell them now but by someone else. Then, like big fish swallowing little fish, larger companies either buy small firms outright or license promising drugs from them. Very often, the original discovery occurs in a university lab with public funding from the National Institutes of Health

Pharma and Profits: Balancing Innovation, Medicine, and Drug Prices, First Edition. John L. LaMattina. © 2022 John Wiley & Sons, Inc. Published 2022 by John Wiley & Sons, Inc.

(NIH), then licensed to a start-up company partly owned by the university and then to a large company. There is very little innovation at the big drug firms."

Dr. Angell went on to say that these "companies are totally dependent on government support" and that "the public should demand something in return for all that government support."

It is quite apparent to anyone involved in R&D that Dr. Angell has no clue about what it takes to discover and develop a new medicine. Her point about the seeds of drug discovery coming from NIH research is, however, absolutely correct. But this is not unique to the biopharmaceutical industry. For decades, Western governments have invested in fundamental technologies in many sectors, including defense, economic growth, nascent industry promotion, and even long-term projects like space exploration. Much of this research is cutting-edge – and a lot of it fails. Yet, when breakthroughs occur, they can change the very order of society, as happened with Covid-19 mRNA vaccines.

Dr. Angell's *Washington Post* piece was not unique. A few years later, the 116th Congress opened with hearings on drug pricing – a bipartisan issue. As the newer members of Congress dug into this, old issues again resurfaced and new outrage was expressed with the role of the NIH in drug discovery touching a big nerve. The following exchange, which occurred between Rep. Alexandra Ocasio-Cortez (AOC) and Professor Aaron Kesselheim (AK) of Harvard Medical School, generated special attention [3].

> AOC: "Would it be correct, Dr. Kesselheim, to characterize the NIH money that is being used in development and research as an early investment? So, the public is acting as an early investor in the production of these drugs. Is the public receiving any sort of direct return on that investment from the highly profitable drugs that are developed from that research?"
>
> AK: "No, in most cases there is - when those products are eventually handed off to a for-profit company, there are not licensing deals that bring money back into the coffers of the NIH. That usually does not happen."
>
> AOC: "So the public is acting as an early investor, putting tons of money in the development of drugs that then become privatized, and then they receive no return on the investment that they have made."
>
> AK: "Right" Source: [3]/American Association for the Advancement of Science.

Such Congressional inquiries are not unique. Back in 2013, Sen. Ron Wyden sent an open letter to NIH Director Dr. Francis Collins urging the NIH to reexamine its policies to protect the interests of taxpayers.

> "The NIH and its mission deserve strong support, yet citizens remain concerned about public dollars being used to research and develop drugs and treatments which are then commercialized with the public getting the short shrift.
>
> Scientific advances rarely happen overnight or as the result of any single agreement – a partnership, CRADA (Cooperative Research and Development

Agreement) or license. But there should be a mechanism in place that ensures that the return on taxpayer investment is considered.

My bottom line: When taxpayer-funded research is commercialized, the public deserves a real return on its investment."

Wyden's letter was triggered by the approval of the rheumatoid arthritis (RA) drug, Xeljanz™ (tofacitinib), which emanated from an early Pfizer–NIH collaboration. Ironically, in the 1980s, Congress was concerned about US industry competitiveness, particularly with respect to the timely transfer of technology from government labs to the private sector. Thus, in 1986 it passed the Federal Technology Transfer Act (FTTA), which allowed government labs like the NIH to enter into Cooperative Research and Development Agreements (CRADAs) with biopharma companies. With CRADAs in place, scientists at the NIH were allowed to conduct collaborative research with industry scientists to promote the discovery of new drugs.

Unfortunately, the collaborations that were expected to boom once the FTTA was passed never materialized. That is because the legislation demanded that the price of any resulting product would have to reflect the taxpayers' investment. In effect, the government would have the right to control the price of any drug that came out of such a collaboration. Naturally, industry avoided such collaborations like the plague.

That was all changed by NIH Director and Nobel Laureate, Dr. Harold Varmus, who recognized the problem and ended this policy. As a result, NIH–industry collaborations began to flourish. This change has helped spur the discovery and development of the drugs now reaching the marketplace: drugs for AIDS, cancer, chronic diseases, and rare diseases – and Xeljanz™. In fact, the story of the discovery of Xeljanz™ is a great example of NIH–industry interaction.

In the summer of 1993, Dr. Paul Changelian, a Harvard-trained chemist then working in Pfizer's immune suppression group, was looking for a new way to prevent organ transplant rejection. Paul, always on the lookout for new ideas, particularly ones related to suppressing an overactive immune system (a hallmark of diseases like psoriasis and RA as well as organ rejection), decided to attend the FASEB Summer Conference on Lymphocytes and Antibodies. This was a meeting held in Vermont, where the world's leading scientists got together to discuss cutting-edge research. Actually, he almost missed the meeting. His wife, Diane, was expecting their second child and her due date was around the same time as the meeting. Paul was hesitant but Diane knew that this conference was important to Paul, so she suggested they both go to Vermont. Their second child could be delivered just as easily in Vermont as in Connecticut, she argued. Thus, they both ventured north.

While at this meeting, Paul ran into an acquaintance, Dr. John O'Shea, a researcher at the NIH. As scientists often do, Paul and John discussed their current projects. When Paul told John that he was hunting for new drug targets that could produce immune suppression, John mentioned that his lab had just discovered a particular enzyme from a class known as kinases that could play a role in immune function. The enzyme is a member of the Janus family of kinases and is commonly known as JAK. JAK was shown to control signaling by interleukin-2, a key growth factor for T-lymphocytes, the cells responsible for the rejection of a transplanted kidney.

Paul was intrigued by these findings, but he did have a concern. At that time, other laboratory data suggested that interleukin-2 was not the only growth factor used by T-lymphocytes, and blocking it via the JAK pathway might not be sufficient to turn down an overactive immune system. The only way to answer this key question was to persuade Pfizer to start a program that might discover a safe and effective JAK inhibitor, then test it in vitro, in animals, in healthy volunteers, and, if all went well, in patients.

Working closely with the NIH, Pfizer was able to sign the necessary agreements to enable the project to begin. But a totally novel project like this was essentially starting from scratch as all of the necessary assays and in vivo models needed to be developed. More importantly, a compound that blocked the activity of the JAK enzyme had to be found, often the most difficult part of such a program. The immunology team worked diligently on this and by 2000, seven years after initiating the program, a JAK inhibitor, CP-690,550 (later to be christened tofacitinib), was nominated for clinical development. Before going to humans, tofacitinib was rigorously studied at Stanford University where animals underwent kidney transplants. Not only was the compound more effective than the standard of care at the time, cyclosporine, but also the tofacitinib-treated transplanted kidneys had no evidence of tissue injury. The researchers had never seen results like this before.

Studies in humans proved equally exciting. A two-week study in patients with severe psoriasis showed that the drug could clear psoriatic plaques almost completely. But the most compelling results were found in RA patients, where tofacitinib worked in patients in whom the currently used biological drugs were ineffective. Pfizer completed clinical trials with tofacitinib in RA and filed a New Drug Application with the FDA late in 2011. The safety and efficacy credentials of tofacitinib were reviewed by the Arthritis Advisory Committee of the FDA and, by an eight to two vote, it recommended approval of this drug for "the treatment of adult patients with moderately to severely active RA." On 6 November 2012, Xeljanz™, Pfizer's brand name for tofacitinib, was approved by the FDA for RA patients.

These patients are the beneficiaries of the hard work and dedication of the hundreds of Pfizer colleagues who worked on this program for over 19 years at a cost of over a billion dollars. But a small degree of thanks should also go to Diane Changelian, who traveled to the woods of Vermont so her favorite scientist could attend a conference, and to their son, Haig, who was kind enough not to arrive before his time.

It is important to ponder the question raised years ago by Eugene Fitzgerald and Andreas Wankerl:

"Are national expenditures on research and development like other government spending or are they a national investment [4]?

   The same question undoubtedly arose 80 years ago as Vannevar Bush and Karl Taylor Compton at MIT foresaw that the use of national resources to fund science and technology research could change the course of warfare. Through a partnership between universities, government and industry, the R&D investments resulted in U.S. defense technology coming of age just in time for World War II. With an even greater vision at the end of the war, Bush portended that such national expenditures in science and technology would have broad, non-defense implications for a national economy. President Obama (in his 2011 State of the Union address) summarized the subsequent success by recalling the space-race investments, with which 'we unleashed a wave of innovation that created new industries and millions of new jobs.' In fact, every nation climbing the global economic ladder has followed the U.S. model" Source: [4]/Forbes Media LLC.

The investment by the United States in basic R&D has fueled growth across other industries beyond defense and space. The basic technology that underpins the iPhone, things like the touch screen and the lithium battery, has its roots in government-funded research. Yet we would not dream of asking Apple to give the American people discounts on their iPhone purchases so "the American people can get a return on their investment." One might argue that the actual return on investment is the tax bill ($38 billion in 2018) that Apple pays every year along with the 80 000 people Apple employs in the United States.

   Unfortunately, many government officials believe that basic research at the NIH is the most crucial part in the development of a new medicine. Yes, that work is important. However, it is the equivalent of running the first mile of a 26.2-mile marathon. You still have 25.2 miles to go and each mile gets more difficult as you progress. In the Xeljanz™ story, an important observation made at the NIH by Dr. John O'Shea led to a CRADA signed with Pfizer scientists. But it took almost two decades, hundreds of scientists, and a billion dollars to make Xeljanz™ a reality for RA patients.

   Decades ago, the United States believed that investments in research were important for the public good – for our defense, for our health, and for our economic prosperity. Furthermore, any progress coming from these investments has to be broadly disseminated and made available so that it can contribute to the common good. Yet, these projects are taken on by industry with great risk. Less than 5% of these programs actually lead to a successful drug.

   If Congress were to reinstitute payback provisions on drugs that emanate from work done by the NIH, we will go back to the situation that existed in the 1980s. Companies will shy away from any contact with programs that can be perceived to have ties with government agencies. At a time when patients are clamoring for new medicines to treat their conditions, at a time when the basic understanding of the causes of diseases are being understood, and at a time when there is a dire need for drugs to treat Alzheimer's disease, drug-resistant infections, various cancers, etc., can we really afford to take a step backwards?

Coming back to the Covid-19 vaccines, it is not surprising that people again claimed that government funded all of the critical work necessary to find mRNA vaccines and that biopharma simply carried out the late stage of the R&D process. Actually, investments by venture capitalists (VCs) really drove the development of these vaccines. Research into the potential use of this technology had been ongoing for decades and progress was slow [5]. It was not until the formation of two biotech companies, Moderna (from "modified RNA") and BioNTech (Biopharmaceutical New Technologies), backed by tens of millions of dollars first from VCs and then by becoming public companies, that the resources were available to make mRNA vaccines a reality. But this did not occur immediately. It took more than a decade before Moderna would yield its first product – a Covid-19 vaccine.

Once again, the wisdom of the late Bob Dole and the impact of the Bayh-Dole Act put this into perspective [6].

> "In just a few short months, the world as we know it has been threatened and transformed by a global pandemic – a pandemic that has already claimed hundreds of thousands of lives and devastated the economies of countries around the globe.
>
> Our nation has faced significant challenges before, and we have always risen to the occasion and prevailed. It's the American way. I remain confident about the resilience in response to the Covid-19 crisis, and much of this confidence arises from legislation that I co-authored in 1980 with my former colleague – the late, great senator from Indiana, Birch Bayh.
>
> The Patent and Trademark Law Amendments Act – commonly known as the Bayh-Dole Act – helped set the stage for the public-private partnerships that are essential to developing a vaccine and effective treatments against the novel coronavirus. Several of the vaccines and therapeutics currently in development likely wouldn't exist without this legislation. According to the Bloomberg Law, 'The most promising Covid-19 treatments and vaccines being explored right now were made possible' because of Bayh-Dole.
>
> This year marks the 40th anniversary of the Bayh-Dole Act. We introduced the legislation because we knew neither government nor industry alone had the intellectual capacity for the resources to develop and produce the kind of innovations that lead to life-saving cures and transformative technologies. As Sen. Bayh and I wrote years ago in a joint op-ed, 'The purpose of our Act was to spur the interaction between public and private research so that patients would receive the benefits of innovative science sooner.'
>
> Many research institutions and universities are responsible for the kind of foundational discoveries and inventions that ultimately lead to innovative new cures and products. But it takes a massive investment and additional research and development by the private sector to bring these innovations to market.

For every dollar the government spends, industry spends 10 to 100 times that amount.

Before Bayh-Dole, the government retained ownership of patents resulting from federally funded research. That meant private firms had no incentive to partner with research institutions or commercialize their inventions. As a result, very few discoveries made it from the lab to market.

Prior to the law, the government licensed just 5% of the 28,000 patents it retained, and few were developed into commercial products. By allowing universities to manage inventions made with government funding, Bayh-Dole paved the way for academic institutions to take the lead in turning their research into real, usable patents – and did so without creating any new bureaucracy or spending taxpayer dollars.

In 1980, few could foresee that our legislation would help spur the development of a children's vaccine for rotavirus, quantum computing, the nicotine patch, FluMis, and transformative companies such as Google. Thanks in part to Bayh-Dole, three new companies are launched and two new products are brought to market every day, on average. The law has also jump-started many small businesses – 70% of university licenses are issued to startups and small companies. To date, the Bayh-Dole Act has bolstered US economic output by $1.7 trillion, supported 5.9 million jobs, and led to more than 13,000 startup companies.

Today, Bayh-Dole is helping facilitate the development of Covid-19 therapies.

For instance, Moderna – a small company in Massachusetts that is about to begin phase 3 clinical trials with its vaccine candidate – counts patent licenses from Harvard University and the University of Pennsylvania in its intellectual property portfolio.

I was humbled when *The Economist* called Bayh-Dole 'possibly the most inspired piece of legislation to be enacted in America over the past half-century. More than anything, this single policy measure helped to reverse America's precipitous slide into industrial irrelevance.'

Sen. Bayh and I partnered in a bipartisan manner to enact this important legislation, and I give him the majority of the credit for his vision and leadership. I am confident that he would join me in urging our present-day leaders to rise above partisan political bickering and work together to defeat this virus. Innovation will be the key, and I remain optimistic and proud that our legacy legislation may play a small role in a victory for millions and millions around the world" Source: [6]/CQ Roll Call.

Moderna, like all biotech startup companies, depended on investors to fund the first decade of its existence. Indeed, the whole biotech ecosystem is based on this funding. Yet, most startups fail. This is not due to incompetence or mismanagement. Rather, early drug R&D is a highly speculative endeavor where, despite great ideas and smart people, hypotheses often fail.

Why, then, do investors plow money into biotech? Well, when the hypothesis is proven, it can be a lucrative investment, one which can benefit millions of people. But such success is based on protection of the intellectual property (IP) that the startup has generated. IP protection defends innovators against predatory competitors who might steal technology and, with lower costs, undercut prices. IP allows innovators to recoup investments and generate profits. Without IP, there would be no investors and the biotech ecosystem would collapse.

As the vaccines began rolling out in 2021, editorials began to appear urging that patent reprieves be granted for the Covid-19 vaccines [7]. The rationale offered for suspending the patent protection for these vaccines was that, by allowing for their widespread production, vaccines would be produced more rapidly thereby ending the pandemic more quickly. But simply dropping IP protection would not automatically boost vaccine supply. Making something as difficult as an mRNA vaccine is a complicated process. This is not like making ibuprofen or acetaminophen pills. The Pfizer manufacturing experience for Comirnaty™ is telling [8].

In April 2020, before Pfizer knew it had a vaccine, Pfizer began the process of developing manufacturing processes to produce the lipid formulation designed to house the mRNA vaccine. These are not just any lipids. A unique combination of four different lipids was needed that could protect the mRNA on the way to cells and then release the mRNA once it got there.

Creating such novel formulations required special machines. These machines, the size of a single-car garage, cost $200 million each and Pfizer bought seven of them for their production plants in Puurs, Belgium, and Kalamazoo, Michigan. But beyond the cost of capital, there were dozens of different ingredients needed to produce the vaccine as well as the "know-how" required for manufacturing. You cannot simply hand over an instruction manual to an organization or country that wants to make its own vaccine and assume that Covid-19 vaccines will be rolling out in a few months. Nor can you refurbish older manufacturing plants and convert them to mRNA vaccine production sites. In fact, estimates are that, even if Pfizer and Moderna did grant patent reprieves for its vaccines, it would be late 2022 when operations were up and running elsewhere. And that assumes that the funds were readily available to finance such operations. As proof of this, Moderna will be spending $500 million to build a new manufacturing plant in Africa to supply its Covid-19 vaccine to this continent. Moderna CEO, Stephane Bancel, said it would take two to four years to complete it [9].

In fact, it turns out that allowing others to manufacturing the Covid-19 vaccines is not needed to inoculate the world. Pfizer will have delivered 3 billion doses of its vaccine in 2021 and another 4 billion in 2022. Moderna, after delivering 1 billion doses of its vaccine in 2021, plans to produce another 3 billion in 2022. That is 11 billion shots of mRNA Covid-19 vaccines – enough to vaccinate nearly 70% of the world's 8 billion people. Adding in vaccines from AstraZeneca, J&J, Novavax, and others, the world should be awash in Covid-19 vaccines by the end of 2022, even if booster shots are needed. The real challenge is establishing the distribution networks needed to get these vaccines to those remote areas that lack the required storage facilities.

If this is the case, why did biopharmaceutical industry react so negatively when the calls were issued for suspending the IP for Covid-19 vaccines? They feared a slippery slope leading to similar demands for any important new drug. That is the death spiral of the biopharmaceutical industry. IP is the foundation of new technologies that benefit everyone. Certainly, this has never been more evident than during the coronavirus pandemic.

Despite its notable manufacturing success, Pfizer has been criticized for how it delivered its vaccine. In May 2021, the website *STAT* published an article entitled "Boxed in: How a single Pfizer decision complicated the Covid vaccine rollout while boosting profits." Authors Olivia Goldhill and Rachel Cohrs claimed that Pfizer's decision to package its vaccine in containers that held 1170 doses was made to maximize profits. They argued that smaller containers with a 100-dose capacity would have been better for delivery to remote locations, but that smaller containers would have been more costly.

These authors claim to be "working to hold corporations and public bodies to account in their response to Covid-19." On this occasion, however, their argument is a stretch. Pfizer's decision on container size was likely made when body bags of Covid-19 victims were being loaded into refrigerator trucks in major cities around the globe. As the pandemic was raging, the priority was to get as many doses to large population areas as soon as possible so that massive inoculations could rapidly occur. The fact that the mRNA vaccines would need to be stored at ultra-cold temperatures also supported the view that densely populated areas were more likely to have the freezers needed to store the vaccine. To imply that Pfizer executives sat around and determined that there would be more profits in using 1000-dose containers is absurd. The large containers made the most sense.

* * * * * * * * * * * * *

The biopharmaceutical industry has a unique role in the discovery and development of new medicines: it proves or disproves medical hypotheses. Nowhere else is this done – not in academia, not at government research labs, and not at research institutes. When Dr. John O'Shea made a key observation with respect to the potential role of an enzyme, JAK, in various immunological conditions, it took the discovery of tofacitinib (Xeljanz™) by Pfizer scientists to demonstrate the utility of O'Shea's seminal observation. What industry does is applied science. Without it, there are no new medicines.

Since the launch of Xeljanz™, a number of other JAK inhibitors have reached the market: Olumiant™, Jyseleoa™, Rinvoq™ all anti-inflammatory drugs for autoimmune diseases; Jakafi™ and Inrebic™ for myelofibrosis and some cancers. More JAK inhibitors are on the way that will treat other inflammatory conditions such as psoriasis, atopic dermatitis, and inflammatory bowel disease. Years of research and billions of dollars have been poured into these research programs, and the emerging drugs will benefit patients around the globe who suffer from these diseases.

The same can be said for the Covid-19 vaccines. Like with JAK, important early work in mRNA research seeded the all-consuming R&D needed to convert this early work into life-saving vaccines. The true value of these vaccines was only shown when the biopharmaceutical industry converted ideas into reality. The rapid execution by the biopharmaceutical industry proved unprecedented – and that proved crucial. The Congressional Budget Office has estimated that Covid-19 will cost $7.6 *trillion* in lost US output in the coming decade [10] – a number that would surely have been higher had the availability of the vaccines been delayed. More critical is, in just the first 6 months of the US vaccination campaign, it is estimated that 140 000 lives had already been saved [11]. This number will grow substantially once the pandemic is finally controlled.

The United States may well spend $100 billion on mRNA and other vaccines before the pandemic ends. But given the trillions that have already been spent on healthcare and economic bailout packages, can we really begrudge paying those biopharma companies that saved us all?

REFERENCES

1. Hopkins, J.S. and Grossman, M. (2021). Pfizer forecasts $33.5 billion in Covid-19 vaccine sales. *Wall Street Journal*. (28 July).
2. Angell, M. (2015). Why do drug companies charge so much? Because they can. *Washington Post* (25 September).
3. Lowe, D. (2019). Rep. Ocasio-Cortez and Where Drugs Come From, In the Pipeline (*Science*). (1 February).
4. Fitzgerald, E. and Walkerl, A. (2011). Why the government needs to invest in innovation. *Forbes.com*. (1 January).
5. Garde, D. and Saltzman, J. (2020). The story of two firms, Pfizer and Moderna, leading the race for approval of a Covid-19 vaccine. *Boston Globe*. (10 November).
6. Dole, B. (2020). Innovation is key to defeating Covid-19. *rollcall.com*. (24 July).
7. Nature Editorial Board (2021). It's time to consider a patent reprieve for Covid vaccines. *Nature* 592: 7. 30 March.
8. Hopkins, J. (2020). How Pfizer delivered a Covid vaccine in record time: crazy deadlines, a Pushy CEO. *Wall Street Journal*. (11 December).
9. Loftus, P. (2021). Moderna to build vaccine – Manufacturing plant in Africa. *Wall Street Journal*. (7 October).
10. Cutler, D.M. and Summers, L.H. (2020). The Covid-19 pandemic and the $16 trillion virus. *JAMA* 325 (15): 1495–1496. 12 October.
11. Gupta, S., Cantor, J., Simon, K.I. et al. (2021). Vaccinations against Covid-19 may have averted up to 140,000 deaths In the United States. *Health Affairs* 40 (19): 1–7.

CHAPTER 5

INSULIN: THE TRUE COST OF A 100-YEAR-OLD DRUG

THERE ARE certainly many issues with our Byzantine health care system, in particular drug availability, cost, and access. It is understandable that the price of drugs has become a hot topic in the United States, with both political parties trying to address this issue. What is surprising, however, is that insulin has become the flashpoint for political outrage. In some ways, insulin has become a symbol of corporate profiteering. Many point out that access to affordable insulin is a life and death matter.

A while back, NBC's *Today Show* reported on the high cost of insulin. The report closed with the comment that, over 100 years ago, the discoverers of insulin, Frederick Banting and Charles Best of the University of Toronto, sold their patent for $1 as they felt that this discovery should be made readily available to all diabetics. Yet new patients are unable to afford their insulin due to high prices. This implication of greed on the part of pharmaceutical companies prompted a look of horror and disgust from the show's host, Savannah Guthrie, a feeling undoubtedly shared by the TV audience. The "$1 patent" anecdote is well-worn, as evidenced by a *New England Journal of Medicine* article entitled: "The U.S. Insulin Crisis – Rationing a Lifesaving Medication Discovered in the 1920s [1]."

Senator Bernie Sanders seized on the issue and took about a dozen people with diabetes on a bus from Detroit to Windsor, Ontario, to get insulin at a pharmacy just minutes from the United States border. Sanders' goal was to show that driving across the border to get one's medication, despite the time and effort involved, was well worth it. By getting their insulin across the border, these patients saved a lot of money [2].

However, this narrative has a major flaw. The insulin purchased in Canada is not the material developed by Banting and Best. Technically, it is not really insulin.

The insulin that Banting and Best first administered in 1922 was isolated from the pancreas of an ox. It was short acting and had to be administered

Pharma and Profits: Balancing Innovation, Medicine, and Drug Prices, First Edition. John L. LaMattina.
© 2022 John Wiley & Sons, Inc. Published 2022 by John Wiley & Sons, Inc.

multiple times a day. Early work by drug companies like Eli Lilly improved the drug's biological profile by formulating it with things like zinc or protamine, leading to longer-acting versions of insulin. Insulin derived from the pancreas of animals would be the mainstay of therapy for the 60 years.

A big breakthrough occurred in the late 1970s when scientists at Genentech prepared human insulin by inserting the human gene for it into *Escherichia coli* bacteria. The result was a significant advance because human insulin is far less allergenic than that extracted from animals. Eli Lilly signed an agreement with Genentech to commercialize recombinant DNA insulin and, in 1982, Lilly brought to market Humulin™ R (rapid acting) and Humulin™ N (intermediate acting).

It turns out that this form of human insulin is very accessible and relatively cheap. In fact, I recently bought a vial of human insulin for $24.88. This was not from the black market, not in Canada, not mail-ordered from overseas. It was 10 minutes away at my local Walmart. I did not even need a prescription. Furthermore, I actually had the option of purchasing three different forms of human insulin, Novolin®, manufactured by Novo Nordisk: Novolin® R (short- to intermediate acting); Novolin® N (intermediate acting); and Novolin® 70/30 (a combination of Novolin® isophane, which is long acting and regular short acting). All are natural human insulins made by recombinant DNA technology. All cost $24.88. Given that a vial usually lasts 7–10 days depending on a patient's needs, the monthly costs for human insulin obtained by Walmart is less than $100 – accessible and affordable – with no real need for rationing.

While this form of human insulin was the basis for treating diabetes for two decades, it is far from perfect. One must rigorously monitor blood glucose to guard against getting it too high or too low. Get the dose wrong and you could end up in the emergency room. This treatment option requires medical supervision [3].

So, given that human insulin is readily available, where is the disconnect? Why is Bernie Sanders bussing people to Canada to get their insulin? When people talk about how expensive their insulin is, they are really not referring to human insulin. They are referring to insulin *analogs*. In the 1990s, drug companies sought to improve upon Mother Nature. Taking injections of human insulin does not truly mimic the physiologic production of insulin generated by one's pancreas. Furthermore, insulin injections can result in a variety of side effects. Could scientists actually synthesize a new version of insulin that would overcome the challenges of injecting the natural hormone? Could scientists make a better drug to treat Type 1 diabetes?

Lilly, Sanofi, and Novo Nordisk, among others, began research programs to seek such an agent. They invested billions of dollars to modify the structure of insulin, creating analogs of the natural product with the hope of finding better drugs. These efforts were successful and yielded new forms of insulin that proved to control diabetes far better than human insulin. These new insulin analogs provide a more protracted biological action of 24 hours, have a smoother time-action profile, and reduce peak action. They show a lack of weight gain and have a significantly lower risk of hypoglycemia (low blood sugar) especially overnight.

These insulin analogs had to go through the standard US Food and Drug Administration (FDA) trials required to demonstrate safety and efficacy, just like any other experimental medicine. As is the norm, these new drugs were patented. And like any other new drug, the companies that developed them priced them to generate significant revenues. But make no mistake; these are new drugs, not the insulin isolated from oxen by Banting and Best. And they are expensive.

Concerns about insulin analog affordability rest on two factors: the list price of the drug and the nature of one's insurance plan. Given the variety of insulin analogs available, manufacturers need to offer deep discounts to gain access to insurance company formularies. List prices for insulin analogs, in fact, have risen dramatically over the past decade. Graph 5.1, depicts the cost over time for the Sanofi insulin analog, Lantus™. The list price of this medicine increased by 126% from 2012 to 2018. However, the net price, the money actually paid to Sanofi, did not rise at all. The net price of Lantus™ in 2018 was essentially identical to that in 2012 [4]. This is an excellent example of how list prices do not reflect the amounts paid by insurance companies, who can expect deep discounts in the form of rebates.

However, despite the fact that Sanofi is not garnering any additional revenue, this list price increase means many patients are spending more for

Graph 5.1   Lantus cost over time. Source: Modified from [4].

their medication. The amount depends on their insurance. Many plans include copayments and these can lead to affordability issues. For example, patients may have to pay the full *list* price of a drug until a deductible is met. Another type of plan requires patients pay a percentage of the drug's list price before a rebate is paid to the insurance company. Either case can result in high out-of-pocket spending for any drug, including insulin analogs. And, of course, those without insurance pay the full freight. Despite the fact that net costs of insulin analogs to payers have decreased thanks to these rebates, these savings have not been passed to the patients. The insurance companies keep the money.

It should be noted that for many privately insured patients, insulin cost-sharing is not overly burdensome. A study from the Harvard Pilgrim Health Care Institute showed mean patient out-of-pocket spending for a 30-day insulin analog prescription was $36 in 2006, peaked at $45 in 2012, and actually fell to $38 in 2017 [5].

This experience, however, is not shared by those without insurance nor those on Medicare or Medicaid, and this accounts for Senator Sanders' bus trips. Canada, like most countries around the world, has a single payer system. Canada negotiates directly with drug companies and there is no "net price/list price" issue, there are no deductible limits. Unfortunately, the United States will not be going to a single payer system anytime soon.

Another player is driving down the cost of insulin analogs for Americans. In June 2021, Walmart began selling its private label version of NovoLog®, the insulin analog manufactured by Novo Nordisk. Designed to benefit those without insurance, Walmart's ReliOn™ NovoLog® insulin vials cost $72.88 (compared to Novartis branded NovoLog®, which costs $289) [6]. A typical patient using two vials of NovoLog® per month will now only spend $1749 per year for Walmart's product as compared to $6945 for brand name NovoLog®.

But an even more important step toward greater access to insulin analogs has been taken by regulators. Historically, the FDA has been slow to approve "biosimilars." These are copies of biological drugs that compete on price. The FDA's hesitancy is based on biochemistry. Small molecule drugs, such as statins or non-steroidal anti-inflammatory drugs like ibuprofen, are manufactured chemicals, which have relatively simple recipes and are relatively easy to copy. Biological drugs, on the other hand, are much larger, more complex molecules made in an engineered living organism such as a genetically modified bacteria or yeast. The FDA feared that subtle differences in the manufacturing process could lead to inexact copies that are not as efficacious as the original medicine, or which might be poorly tolerated due to immunogenicity. As a result of this hesitancy, insulin analogs, whose patents have been long expired, had no competition from generic forms, i.e. biosimilars [7].

This all changed when the FDA issued a new policy that created a streamlined process to permit approved biosimilar insulin analogs to be interchangeable with brand name analogs so that they could be dispensed in place of those products without the need for a new prescription. With this new guidance in hand, companies set out to develop insulin analog biosimilars. On 29 July 2021, the first of these, Semglee™, from Viatris and Biocon, was approved by the FDA [8]. Semglee™ is a biosimilar to Sanofi's product Lantus™ and shows no differences in terms of clinical efficacy, safety, purity, and potency. Viatris and Biocon will have 12 months of marketing exclusivity before the FDA can approve another interchangeable version of Lantus™, but it is anticipated that others will follow. Semglee's™ price was not immediately announced, but Walmart's ReliOn™ NovoLog® shows what is possible.

The efforts by Walmart and Viatris/Biocon are not panaceas. Ideally, access to insulin analogs should be universal and affordable in every health-care plan. But certainly, the situation is improving as we move to affordable insulins.

Finally, President Biden's Build Back Better social spending plan also seeks to address the cost of insulin analogs [9]. A provision in the agreement would cap the price of these drugs for diabetics at $35/month. Whether this Bill actually gets passed and whether this provision of the Bill actually survives remains to be seen. However, it is entirely possible that at some point in the future some form of a price cap on insulin analogs could be enacted.

* * * * * * * * * * * * * *

We have come a long way from injecting ox pancreas extract to treat diabetes. Years of research have been invested and billions have been spent, first on generating human insulin and then in developing modified insulins that are significantly better treatments. Controlling one's blood sugar today requires less frequent injections and these new drugs' more reliable delivery profiles reduce dangerous hypoglycemic events. In addition, the weight gain seen with insulin does not occur. The scientists at the drug companies that developed these insulin analogs have indeed improved on Mother Nature and, in doing so, have created new medicines that provide real value. Finally, we are reaching a point where the prices of insulin analogs will be affordable for all. No longer do diabetics have to go to Canada for their medicines. Walmart and biosimilars offer better solutions.

REFERENCES

1. Fralick, M. and Kesselheim, A.S. (2019). The U.S. Insulin Crisis – Rationing a Lifesaving Medication Discovered in the 1920, *New England Journal of Medicine* 381: 1793–1795. 7 November.

2. Ember, S. (2019). Bernie Sanders Heads to Canada for affordable insulin. *New York Times* (28 July).
3. Tribble, S.J. (2015) You can buy insulin without a prescription, but should you? *KHN News* (14 December).
4. Sanofi (2019). Pricing Report: Advancing Responsible Leadership. https://www.sanofi.us/-/media/Project/One-Sanofi-Web/Websites/North-America/Sanofi-US-Home/corporateresponsibility/Prescription_Medicine_Pricing-2019.pdf.
5. Meiri, A.; Zhang, F.; Ross-Degnan, D. Wharam, J.F. (2020) Trends in insulin out-of-pocket costs and reimbursement price among U.S. patients with private health insurance, 2006–2017, *JAMA Internal Medicine*; 180 (7) 1010–1012, 1 July.
6. Molina, B. (2021) Walmart launching its own low-cost insulin to revolutionize affordability for diabetics, *USA Today*, June 29.
7. Socal, M.P. and Greene, J.A. (2020). Interchangeable insulins – New pathways for safe, effective, affordable diabetes therapy. *New England Journal of Medicine* 382 (11): 981–983. 12 March.
8. Brennan, Z. (2021). Insulin biosimilar wins first-ever interchangeable designation from FDA. *Endpoint News* (29 July).
9. Wise, A. (2021). Democrats' deal on prescription drugs would lower costs for seniors. *NPR*. https://www.npr.org/2021/11/02/1051649532/democrats-deal-on-prescription-drugs-would-lower-costs-for-seniors (2 November).

CHAPTER **6**

THE COSTLY ALZHEIMER'S DISEASE DRUG: A QUESTIONABLE BREAKTHROUGH

Aт THE dawn of Trump's presidency, there were rumblings that the new government would eliminate a lot of bureaucracy that slows innovation. Most applauded this initiative. After all, who likes red tape? But there were also concerns. The cautious procedures of the Food and Drug Administration (FDA) are there for a reason – to protect Americans. The implied threat to rigorous drug approval prompted me to write the following editorial for *Science* magazine [1].

> As the Trump administration takes shape, there is much speculation as to what major changes will be made. A dominant theme of the Trump campaign was to cut through bureaucracy in Washington, D.C., thereby enhancing innovation and bringing new technology to Americans more quickly. Nowhere could such a philosophical change have more impact than on Food and Drug Administration (FDA) approval of new drugs.
>
> Decades ago, it was not unusual for the FDA to approve a drug simply on the basis of its effect on disease biomarkers. For example, approval of the first statins, including Merck's Mevacor™ and Zocor™, was based on their ability to lower low-density lipoprotein (LDL) cholesterol, known to be associated with the buildup of deadly atherosclerotic plaque. It wasn't until many years later that long-term clinical trials involving thousands of patients actually proved that these drugs reduced heart attacks and strokes.
>
> However, there are far more examples of drugs that failed to provide the desired health outcome despite having favorable effects on disease biomarkers. One example comes from another class of lipid modulators, the cholesteryl ester transfer protein inhibitors (CETPis). When discovered, these drugs looked promising in that they raised high-density lipoprotein (HDL) cholesterol, long associated with cardioprotection, as well as lowered LDL cholesterol,

Pharma and Profits: Balancing Innovation, Medicine, and Drug Prices, First Edition. John L. LaMattina.
© 2022 John Wiley & Sons, Inc. Published 2022 by John Wiley & Sons, Inc.

thereby completely remodeling the lipoprotein profile of patients with heart disease. Despite their promise, the FDA would not approve a CETPi without evidence that it did in fact reduce adverse cardiovascular outcomes. To date, clinical trials for these drugs have shown that they do not reduce heart attacks and, in one case, actually increased deaths.

Such occurrences are not exclusive to the cardiovascular field. For Alzheimer's disease (AD), several companies have discovered experimental drugs that reduce the level of amyloid proteins, which are believed to be an initiating factor in AD. However, despite successfully lowering amyloid levels in AD patients, long-term studies showed that these drugs do not improve cognition in this population. These examples are not unique. There are cases where drugs that lower plasma glucose do not prevent the adverse health consequences of diabetes, and where drugs that shrink tumors do not reduce cancer deaths. Favorably affecting disease biomarkers is an important step in drug research and development (R&D) because it indicates that the experimental drug could potentially prevent the long-term consequences of disease. However, only long-term clinical trials, although time-consuming and expensive, can prove the value of a new medicine.

President Trump and Tom Price, the newly sworn-in Secretary of Health and Human Services (which oversees the FDA) want to streamline the FDA's approval process for drugs. But changing the current drug approval paradigm would hurt everyone. What if the FDA simply rules that early-stage drugs are safe solely on the basis of biomarker effects? Suppose doctors and patients decide on whether to take such a drug with unproven long-term effects? This would certainly get experimental drugs to patients faster and enable companies to recoup their R&D investment more quickly. However, based on historical precedent, many of the drugs approved in this way would be ineffective. At best, they would be glorified placebos. At worst, they could prove harmful. As a result, patients would not have benefited by taking such a drug, physicians would be called into question for prescribing such a drug, payers would have wasted millions (even billions) of dollars reimbursing the costs of such a drug, and companies would face the inevitable lawsuits that arise in these situations. Everyone loses.

Over the years, the FDA has learned that, for many diseases, it is important that the drug maker prove the benefits of the drug before unleashing it on the public. Does this slow the process and greatly add to the R&D costs? Absolutely. But it is good medical practice Source: [1]/American Association for the Advancement of Science.

This piece was written out of concerns that the FDA would, due to political pressure, approve new drugs without sufficient understanding of their full profile, particularly with respect to risks versus benefits. It never occurred to me that the FDA, without government influence, would approve a drug for Alzheimer's disease (AD) based predominantly on biomarker data.

Over the past 30 years, tremendous progress has been made in treating cancer, heart disease, diabetes, arthritis, and more. But one major area has seen little progress – AD. A few drugs, such as Aricept™ (donepezil), have made it to patients but these are palliative treatments that only slow the onset of the disease by a few years at best. Many drug candidates have been taken into human studies but they have uniformly failed to reverse the course of disease. Brain research is very difficult.

Since the 1980s, the AD research field has been dominated by what is known as the amyloid hypothesis. In 1906, German neuropathologist Alois Alzheimer found that the plaques found inside the brain of a deceased dementia patient were made up of sticky protein deposits. Such deposits were ubiquitous in AD patients. By 1984, scientists had identified beta-amyloid – the protein fragment that makes up these plaques. It was believed that beta-amyloid deposits were killing off neurons thereby leading to AD. Further, it was theorized that removal of beta-amyloid plaques should stop or slow the progression of AD. The race to find drugs was on.

But the race turned into a slog. Various antibodies were studied for their ability to remove beta-amyloid from the brain. Typical was the experience with bapineuzumab, a monoclonal antibody that targeted beta-amyloid oligomers and plaques. In human studies, bapineuzumab significantly reduced an AD biomarker in the cerebrospinal fluid (CSF), which boded well for success in treating this disease. There were, however, patients treated who showed no improvements in either cognition or brain function. Despite the biomarker data, the patients received no benefit from this drug [2]. This experience was repeated for other AD drugs taken into human trials.

There was a last chance to prove the amyloid hypothesis – Biogen's aducanumab. Biogen (along with its partner in this endeavor, Eisai) believed that aducanumab would succeed where others had failed. Their data showed it was more potent and removed more beta-amyloid than previous entrants in this field. But, in March of 2019, Biogen halted AD trials for aducanumab. The independent monitoring committee for the phase 3 studies had done an interim analysis and concluded that this drug was unlikely to show any benefit compared to placebo [3]. Biogen chief executive officer (CEO) Michel Vounatsos delivered the gloomy results:

"This disappointing news confirms the complexity of treating Alzheimer's disease and the need to further advance knowledge in neuroscience. We are incredibly grateful to all of the Alzheimer's disease patients, their families and the investigators who participated in the trials and contributed greatly to this research" Source: [3]/STAT.

Why did all these drugs fail despite the impressive biomarker data that were generated? First, there are different isoforms of beta-amyloid and no one antibody targeted them all. Second, AD is a slowly evolving disease and deciding what types of AD patients to study is not obvious. For example, early studies looked at patients with advanced disease, but it appeared that the disease had progressed too far in these patients for the drugs to be effective. But challenges exist even when running studies in patients with mild or moderate disease. To detect true disease-modifying effects of the experimental drug, trials need to run long-term and include large numbers of patients (and even greater numbers of clinical sites). These types of clinical trials are subjected to placebo responses, and loss of subjects and life-events, all of which can undermine the value of data. AD is complicated. AD research is challenging.

When Biogen stopped these studies, it appeared the amyloid hypothesis was dead. Surprisingly, there were some in the research community who were not disappointed. They claimed that dogmatic beta-amyloid advocates had sequestered most of the AD research funding. Some believed that the stifling of other ideas was the reason there were no viable treatments for AD [4].

However, such second guessing was short lived. In a major reversal, just seven months after it halted aducanumab studies, Biogen announced that it would indeed pursue regulatory approval for its drug [5]. What changed? Biogen conducted a new analysis of the data from its two trials, known as EMERGE and ENGAGE. When these trials were stopped in March, Biogen had data from 1748 patients who had completed 18 months of treatment. Some months later, a larger dataset of 3285 patients became available including another 2066 who had also completed 18 months in the study.

The ENGAGE study did not meet its primary endpoint. However, Biogen claimed that the EMERGE study did meet its primary endpoint. Decline slowed by 22%, versus placebo ($p = 0.01$) in a standard scale used to measure dementia.

Essentially, one study failed but one worked, albeit modestly. Normally, the FDA requires two successful phase 3 studies to justify approval of a new drug, but given the dire need for drugs to treat AD, Biogen decided to push forward and file a New Drug Application (NDA).

As part of its review process, the FDA convened its Peripheral and Central Nervous System Drug Advisory Committee to review and opine on the aducanumab data package. The main question asked of the committee was: "*Does Study 302 (EMERGE), viewed independently and without regard for Study 301 (ENGAGE), provide strong evidence that supports the effectiveness of aducanumab for the treatment of Alzheimer's disease?*" (It was clear from this question that the FDA felt that the ENGAGE study was not successful and was going to base its ultimate decision on the EMERGE study.) Eight members said "no." Two others were uncertain. Only one voted "yes [6]." In the vast majority of cases, the FDA follows the advice of its expert advisors. But this time was different.

The FDA took an unusually long time to reach its decision. But on 7 June 2021, it shocked the world by approving the drug [7]. As would naturally be expected, Biogen was thrilled by the decision as seen in this statement from Biogen CEO Vounatsos [8]:

> "This historic moment is the culmination of more than a decade of groundbreaking research in the complex field of Alzheimer's disease. We believe that this first-in-class medicine will transform the treatment of people living with Alzheimer's disease and spark continuous innovation in the years to come"
> Source: [8]/STAT.

Biogen was elated. Many others were outraged. In a very unusual move, the seven advisory committee members who voted against approval wrote an editorial in the *New England Journal of Medicine* [9] decrying the FDA decision. The gist of their opposition was that the FDA approved the drug on the basis of its ability to lower beta-amyloid. The authors argued, however, that there is not a strong scientific basis for using beta-amyloid as a reasonable surrogate for AD. They pointed out that more than two dozen therapies based on the amyloid hypothesis had already gone through late-stage clinical trials and none showed a meaningful clinical benefit.

In fact, after the FDA's approval, three members of the advisory committee that rejected aducanumab resigned from the committee [10]. One was Dr. Aaron Kesselheim, professor of Medicine at Harvard Medical School, who said that the FDA decision was wrong "because of so many different factors, starting from the fact that there is no good evidence that the drug works." Dr. Joel Perlmutter, a neurologist at Washington School of Medicine in Saint Louis, also resigned saying that the FDA's claim that the drug's ability to attack the beta-amyloid protein in a patient's brain would slow cognition decline was "a major problem."

Not everyone was displeased. Dr. Gil D. Rabinovici, professor of Neurology at University of California San Francisco, wrote in the *New England Journal of Medicine* [11]:

> "Many clinicians, researchers and patient advocates lauded the approval, asserting that even modest slowing of disease progression by aducanumab would translate into meaningful benefit and provide hope for the estimated 6.2 million Americans with Alzheimer's disease."

However, a poll conducted shortly after the drug's approval indicated little support in the medical community [12]. Of 200 primary care physicians and neurologists polled, 87% either disagreed or strongly disagreed with the FDA's decision. Another issue for these doctors was that the FDA approved a broad label for aducanumab allowing it to be prescribed for any patient with cognitive impairment, not just AD. Yet, the Biogen study that showed lowering of beta-amyloid was carried out in patients with mild AD. Thus, only 47% of the neurologists and 34% of the primary care doctors believe that they had enough information to make a prescribing decision. But that might be moot. More than 70% of these physicians said that they would *not* prescribe aducanumab.

It should be noted that as part of its approval for aducanumab, Biogen has committed to a phase 4 study to show the reduction of beta-amyloid actually correlates with a slowing of cognitive decline. But the results will not be available anytime soon. Biogen has 10 years to run the study.

At FDA approval, Biogen announced the brand name for aducanumab – Aduhelm™. Biogen also announced its price – $56 000/patient for a year's therapy. That is the list price, not the net price nor the co-pay that patients with insurance will pay (depending on the type of coverage). The Institute for Clinical and Economic Review (ICER), a nonprofit organization that measures the value of new medicines, provided cost-effectiveness estimates for Aduhelm™ of between $2500 and $8300/year [13]. Part of ICER's challenge was trying to figure out how much benefit the drug actually provides. If the drug does, in fact, delay cognitive decline by 22% – or about 4 months over an 18-month period – is that worth $56 000/year? That is a question that health insurers wrestle with.

The safety of Aduhelm™ is another important consideration. Pooling of data from both the ENGAGE and EMERGE studies identified brain swelling or tiny hemorrhages, called amyloid-related imaging abnormalities (ARIA), as the major adverse effect [14]. Of the 1105 patients treated with the high dose, 454 experienced ARIA. Of those, 76% were asymptomatic and, of those reporting symptoms, the majority were mild or moderate. Only five patients had severe symptoms but these are far from trivial [15]. However, to guard against ARIA, patients need to get periodic magnetic resonance images (MRIs), which add to the cost of therapy.

It is no surprise that, given the FDA's approval for Aduhelm™ based solely on lowering of beta-amyloid, other potential AD drugs are being rapidly advanced. Lilly has decided to file an NDA for its drug donanemab, which also clears beta-amyloid from the brain. Lilly believes that its drug is more effective than any other treatment [16]. In addition, Biogen and Eisai are seeking another fast approval for a second antibody, lecanemab, that also lowers beta-amyloid in the brain. Lecanemab has been touted as a next-generation agent and the NDA filing also will be based on the drug's impact on biomarkers [17]. Given these new relaxed standards by the FDA, it would not be a surprise to see more companies seeking drug approvals without data showing actual cognitive function improvements. It is also hard to comprehend why the FDA would suddenly raise the bar for future entrants. The barn door is already wide open.

Even though we will not have the results for the Aduhelm™ phase 4 study for a decade, there is a lot riding on it. But what if the trial fails?

That is not necessarily a remote possibility. Drugs that act on the central nervous system (CNS) are notorious for giving false positive results due to large placebo effects. In the case of Aduhelm™, the key outcome in patients is better cognitive effects in those who received the drug as compared to those who got a placebo. But, patients on placebo always show some benefit in CNS trials complicating such studies. It will be crucial for Biogen to show a more profound and statistically significant effect for Aduhelm™ over placebo. A positive outcome is certainly not a given and these results will be anxiously awaited.

Failure of Aduhelm™ in this study will have profound consequences, most notably for patients. Aduhelm™ provides hope for those in the early throes of AD. Failure will presumably result in the FDA pulling the drug from the market. That would mean that those who were on this drug for years would have been exposed to essentially a placebo – but not one that is totally benign given that ARIA has been found in patients on the drug. A much bigger issue will be dealing with those patients who believed that they have benefitted from the drug (remember – placebo effect). Will the FDA allow them to continue receiving the drug? If so, will Biogen continue to provide it? Who would pay for Aduhelm™ in such a scenario?

A second set of consequences involves the costs to the healthcare system given Biogen's original list price for Aduhelm™ of $56 000. It is probable that thousands of people in the early stages of AD will be treated with this drug in the next decade. If Aduhelm's™ efficacy is shown to be ephemeral, some billions of dollars will have been spent on a flawed drug. And that does not include all the ancillary costs generated such as the positron emission tomography (PET) scans needed to prove the presence of amyloid in the brains of patients (a necessary step before a patient is qualified to take the drug) along with other needed care in the monthly administration of Aduhelm™. Some billions of dollars might be spent on Aduhelm™ before the results of Biogen's new trial are available.

But, the reputation of the FDA is also on the line. The FDA's own advisory committee voted *against* approving Aduhelm™ at its review last summer. It is rare for the FDA to approve a drug when such a negative position is taken by its advisors. One has to be sympathetic with the FDA's position.

They have been battered by patients and advocacy groups (including the Alzheimer's Association) to approve this drug as AD sufferers and their families are desperate for any promising treatment. However, should the Biogen confirmatory study fail, there will be many who will attack the FDA for lowering its standards. The very body that the United States expects to protect it from ineffective medicines will have failed.

Finally, the entire biopharmaceutical industry will be tarnished if Aduhelm™ needs to be discontinued. Despite the goodwill generated by the industry's stellar performance in battling Covid-19, all will be undone in this scenario. Industry critics will seize on such an event as an example of the FDA being too cozy with biopharma and not being more rigorous in its review of new drugs. Furthermore, critics will also focus on the price of Aduhelm™, using it as an example of a greedy industry that needs to be brought under control. We have heard these attacks before.

There is a lot riding on Biogen's phase 4 study for Aduhelm™. For patients, payers, regulators, and those working in the biopharma industry, let us hope for positive data.

In the meantime, while we begin the long wait for these results, Aduhelm™ sales are not taking off. Major institutions like Mass General Brigham [18] and the Cleveland Clinic [19] will not offer this drug to patients. Biogen sales representatives have been banned from detailing Aduhelm™ in neurology clinics like the Neurology Center in Washington, D.C. [20]. Even the US Department of Veteran's Affairs has decided not to include Aduhelm™ on its formulary. Its Pharmacy Benefit Manager cited the "lack of evidence of a robust and meaningful benefit and the known safety signal [21]". These and other rejections by healthcare plans around the country led to estimates that less than 100 patients had been dosed with the drug in the first 2 months after it had been approved [22]. This is stunning given the major need for an effective AD drug.

Given this situation, Biogen took the extraordinary step of cutting the list price of Aduhelm™ by almost 50 percent to $28 200 [23]. Whether this helps to improve sales remains to be seen. For critics, however, given that the drug has not yet been proven to be effective, this move was not embraced. Why prescribe Aduhelm™ now just because the price has been cut in half? The issues with this drug despite the price reduction remain.

* * * * * * * * * * * * * *

For someone like myself, a strong supporter of the biopharmaceutical industry (which I have been a part of for over 45 years), the whole Aduhelm™ situation has been tough to watch. AD is perhaps our biggest medical and social challenge and we desperately need new drugs to treat it. Biogen scientists have worked for years trying to find a treatment. But the fruit of their efforts, Aduhelm™, is not being met with hosannas. Unfortunately, the FDA has approved a drug without having the necessary data to allow patients and physicians to be convinced that this is the answer to their prayers. Without such data, it is hard to support the use of Aduhelm™. What I warned against in 2017 happened four years later. The FDA unleashed a medicine on the American public without requiring the drug maker to prove its benefits. And that is a tragedy.

REFERENCES

1. LaMattina, J.L. (2017). Approving new drugs, *Science* 355, 777, 24 February.
2. Mehta, D.; Robert Jackson, Gaurav Paul, Jiong Shi, Marwan Sabbagh (2017). Why do trials for Alzheimer's disease drugs keep failing? A discontinued drug perspective for 2010–2015. *Expert Opinion on Investigational Drugs* 26(6) 735–739, 1 June.
3. Feuerstein, A. (2019). Biogen halts studies of closely watched Alzheimer's drug, a blow to hopes for new treatment. *STAT* (21 March).
4. Begley, S. (2019). The maddening saga of how an Alzheimer's 'cabal' thwarted progress toward a cure for decades. *STAT* (25 June).
5. George, J. (2019). Biogen Reverses Take on Aducanumab Trial Data. *MedPage Today* (22 October).
6. Herper, M., Feuerstein, A., and Garde, D. (2020). Expert Panel votes down Biogen's Alzheimer's drug, and rebukes the FDA in the process. *STAT* (6 November).
7. Beluck, P. and Robbins, R. (2021). F.D.A. Approves Alzheimer's Drug Despite Fierce Debate Over Whether It Works. *New York Times* (7 June).
8. Feuerstein, A. and Garde, D. (2021). FDA grants historic approval to Alzheimer's drug designed to slow cognitive decline. *STAT* (7 June).
9. Alexander, G.C., Knopman, D.S., Emerson, S.S. et al. (2021). Revisiting FDA Approval of Aducanumab. *New England Journal of Medicine* 385 (9): 769–771. 26 August.
10. Belluck, P; Robbins, R. (2021). Three F.D.A. Advisers Resign Over Agency's Approval of Alzheimer's Drug. *New York Times* (10 June).
11. Rabinovici, G.D. (2021). Controversy and Progress in Alzheimer's Disease – FDA Approval of Aducanumab. *New England Journal of Medicine* 385 (9): 771–773. 26 August.
12. Silverman, E. (2021). Poll: Most neurologists and primary care doctors disagree with FDA's approval of Alzheimer's Drug. *STAT* (30 June).
13. Silverman, E. (2021). Revised analysis nudges up cost-effective price for Biogen's Alzheimer's drug to $3,000 to $8,400 a year. *STAT* (30 June).
14. Hoffman, M. (2021). ARIA Events Most Often Asymptomatic in Aducanumab Treatment. *Neurology Live* (26 July).
15. Saltzman, J. (2021). Reetired neurologist with Alzheimer's knows firsthand the risk of Biogen's new drug. *Boston Globe* (31 July).
16. Mishra, M. and O'Donnell, C. (2021). Eli Lilly bets on Alzheimer's disease drug data as it chases Biogen. *Reuters* (3 August).
17. Feuerstein, A. (2021). Biogen, Eisai seek fast approval for second Alzheimer's drug relying on relaxed FDA standard. *STAT* (28 September).

18. Saltzman, J. (2021). Mass General, Brigham won't offer Biogen's new Alzheimer's drug. *Boston Globe* (28 September).
19. Belluck, P. (2021). Cleveland Clinic Says It Won't Administer Aduhelm™ to Patients. *New York Times* (14 July).
20. Higgins-Dunn, N. (2021). Do not enter: Biogen reps banned from D.C.-area neurology clinics over controversial Alzheimer's drug Aduhelm™. *Fierce Pharma* (22 September).
21. Brennan, Z. (2021). VA decides against adding Biogen's Aduhelm™ to its formulary as PBM shuns controversial Alzheimer's drug. *Endpoint News* (11 August).
22. Joseph, A. Cohrs, R. (2021). Alzheimer's patients are in limbo as hospitals, insurers grapple with whether to offer Aduhelm™. *STAT* (4 August).
23. Belluck, P. (2021). Biogen Slashes Price of Alzheimer's Drug Aduhelm™, as It Faces Obstacles. *New York Times* (20 December).

CHAPTER 7

GENE THERAPY: HOW MUCH IS A LIFE WORTH?

THE ANNUAL JP Morgan Healthcare Conference brings together the world's leading innovators to discuss today's health and wellness challenges. Held each January in San Francisco, thousands of scientists and entrepreneurs attend not just to learn, but also to share ideas and perhaps seek investment opportunities in emerging technologies. When I attended some years back, I had an interesting encounter with a cab driver who took me from the airport to my hotel. When he learned I was attending the conference, he expounded on some of his own medical theories. The most startling was his assertion that drug companies actually had already found a number of cures for diseases like cancer, diabetes, arthritis, etc. However, drug companies kept these cures hidden from the public. His explanation for such absurd behavior was that, if these companies actually cured diseases, they would not make any money. Drug companies thrive on the money they make by having patients taking their pills every day for the rest of their lives, he said.

I exited the cab as quickly as I was able. However, this cab driver did provide an interesting thought. If a company was, in fact, able to cure a disease like diabetes, how would that treatment be priced? Are there profits to be made in a "one and done" business model? Well, this is no longer science fiction. We are now at the dawn of gene therapies, a time when diseases actually can be cured.

It is a diagnosis that no parent wants to hear – their child has acute lymphoblastic leukemia (ALL). While a treatable cancer of the bone marrow and blood, traditional drug treatment regimens can be brutal and not always successful. But, in August of 2017 the US Food and Drug Administration (FDA) announced the approval of a new gene therapy to treat ALL. Known as Kymriah™ (tisagenlecleucel), this Novartis product was the first gene therapy approved in the United States, thereby opening a new era in medicine.

The value of Kymriah™ was wonderfully captured by Dr. Lisa Rosenbaum in the *New England Journal of Medicine*, where she outlined the

Pharma and Profits: Balancing Innovation, Medicine, and Drug Prices, First Edition. John L. LaMattina.
© 2022 John Wiley & Sons, Inc. Published 2022 by John Wiley & Sons, Inc.

development of chimeric antigen receptor T-cell (CAR-T) technology – the science on which Kymriah™ is based [1]. Dr. Rosenbaum, however, did more than talk about the underlying science. She put this research into context with the story of Emily Whitehead who, at the age of five, was diagnosed with ALL. Emily was put through the rigors of multiple rounds of chemotherapy to no avail. Emily's physicians told her parents to consider hospice, but, instead, they entered her in a clinical trial for Novartis' Kymriah™. The procedure worked and Emily was cured of ALL.

Kymriah™ is not a pill. CAR-T therapy is based on removing T-cells from a child's blood, genetically modifying them to respond to specific targets on the child's cancer cells, and then reinfusing these modified cells. These new T-cells then seek out and destroy cancer cells. The results on which Kymriah™ was approved were pretty striking. Of the 63 evaluable children who received Kymriah™, 83% had complete elimination of malignant cells at three months.

Rosenbaum was pretty balanced in her appraisal of the early work with Kymriah™. CAR-T therapy can cause a severe side-effect known as cytokine release syndrome and so this treatment needs to be carried out at appropriate medical institutions that are prepared to handle not just cell reinfusions but also any subsequent issues.

But the launch of curative gene-based therapies brought along another concern – drug pricing. How much is such a treatment worth? For Emily Whitehead's parents, Kymriah's™ value is priceless. It saved their daughter. But not all share that view.

As Kymriah™ neared FDA approval, many analysts speculated that Novartis would charge anywhere between $600 000 and $750 000. Instead, Novartis set the price at $475 000. Furthermore, Novartis took the stance that payment would not be made unless the patients had responded to Kymriah™ at the end of the first month after treatment. Nevertheless, critics assailed Novartis for this price. Not surprisingly, payers were first to weigh in. Outspoken biopharma critic Steve Miller, then of Express Scripts, claimed the price is "dramatically higher than other specialty drugs." Actually, this statement was not true. There were, at that time, drugs that cost twice as much. Kymriah™ barely made the top 10. But Miller's posturing was to be expected. After all, Express Scripts and other payers would be facing the approval of other gene therapies in the coming years.

It is important to look at the arguments used by industry critics who attack Kymriah's™ price. Unfortunately, they start by using the tired argument that taxpayers have funded the early research on CAR-T therapy largely through funding the National Institutes of Health (NIH). One can certainly argue that it was NIH funding that sired the technology leading to CAR-T therapy. However, as was discussed in Chapter 4, this can be said of many technological breakthroughs, not just those in medicine.

The next challenge put forward by critics is around the costs incurred by Novartis in developing Kymriah™ and bringing it to market. They doubt Novartis' claims that it took $1 billion to achieve this. They also challenge the costs per perfusion put forth by the manufacturer. Essentially, critics use their arguments to challenge the amount of profits that Novartis, or any company for that matter, is justified to reap for a life-saving medication. They feel that a price of $160 000 is more than enough to recompense Novartis.

The price of any drug should not be based on the amount of money that was required to develop, manufacture, and sell an individual new medicine. The price should be based on the *value* the drug brings to the patient and the healthcare system. This is rarely discussed. Kymriah™ restores the life of a child or teenager, which is invaluable to his or her parents and family. In addition, the contributions that a person makes to society over the course of a lifetime are significant. How do we value that?

Hard discussions are needed to determine what constitutes fair and reasonable pricing. Core to such a discussion is the value that is placed on every good year of life that a patient gains on treatments like gene therapies. What is an extra year of life worth? Actually, some countries have already made such determinations, using a metric known as "quality-adjusted life year" (QALY). A QALY is used to help set the price in countries with single-payer systems. For example, the United Kingdom rates a QALY at approximately $50 000.

In the United States, prices are not set by the government and so the concept of using QALY as a measure of cost-effectiveness for new drugs has not yet prevailed. However, government agencies like the Centers for Medicare & Medicaid Services (CMS) as well as the FDA do make use of cost–benefit analyses. By executive order, federal agencies must analyze costs and benefits when issuing regulations. Significant new rules only are adopted when the benefits outweigh the costs. In 2011, when the FDA added new requirements for the display of health warnings on cigarette packages and in cigarette advertisements, the agency estimated that annual healthcare savings of $185–601 million would be realized [2]. In making this calculation, the FDA used $200 000 per QALY. In 2015, the FDA established minimum standards for the safe growing, harvesting, packing, and holding of produce to minimize adverse health consequences or death from the consumption of contaminated produce. The calculated savings were roughly $1 billion per year, based on $459 900 per QALY. In fact, looking at the significant rules issued by CMS and the FDA, the associated cost–benefit analyses used a median value of $293 000 and $300 000, respectively, for an incremental year of life.

Thus, as life-saving gene therapies are developed, it would appear that costs of these treatments on the order of $200 000–300 000 per life-year could be justified using health agencies' own analyses. It is unlikely that, when justifying these changes, the FDA thought its QALY figures could also be applied to demonstrate the value of gene therapies. But it would be strange to employ one standard for certain decisions and then turn around and use a lower QALY figure for life-saving procedures when both types of decisions have important effects on the public's health.

Novartis has been exploring other uses of Kymriah™. This therapy has been shown to help non-Hodgkin lymphoma (NHL) patients who have failed two lines of treatment and the five-year durability data are impressive. Despite its $475 000 price tab, Kymriah™ posted sales of $298 million for the first half of 2021 [3]. The families of those who had ALL and NHL and who benefited from Kymriah™ believe this is money well spent.

The Kymriah™ price certainly raised eyebrows, but it appears quaint when compared to Novartis' next gene therapy.

At its annual R&D investor day in November 2018, Novartis updated the world on its drug pipeline and research priorities. It was an impressive session befitting a company that at the time spent $9 billion on R&D (about 18% of its top line revenues). Novartis touted a pipeline that included more than 200 programs in clinical development encompassing over 500 ongoing clinical trials with 26 potential blockbusters in late-stage development. Its pipeline was truly enviable and teemed with potential new treatments in a variety of therapeutic areas including various types of cancer, heart disease, metabolic disorders, and rare diseases.

Often at a session like this, a company will highlight specific programs that set it apart from the competition. Novartis certainly did that with a few of its drugs. But it paid special attention to one program: AVXS-101 – a gene therapy to treat spinal muscular atrophy (SMA). SMA is a rare and inherited neurodegenerative disease. Babies that inherit the most fatal form of this disease do not live past the age of two.

AVXS-101 was developed by the biotech company, AveXis. It is a gene therapy that, after a single dose, corrects the genetic defect that causes SMA. Based on the positive clinical data generated by AveXis from a small clinical trial in children, Novartis acquired AveXis for $8.7 billion to boost its gene therapy portfolio.

Given the price paid for AveXis as well as the commitment that Novartis is making to gene therapy research, it is no surprise that the AVXS-101 program was highlighted at its R&D day. Novartis focused on the drug's exciting life-saving potential, the powerful technology it developed to deliver not just this therapy but also others for central nervous system (CNS) diseases, and the other therapies that can be generated by the AveXis technology platform.

But, despite this impressive array of breakthroughs, the slide that garnered the most attention from the AveXis part of the presentation was the one that dealt with price. The slide compared the value of AVXS-10 in curing a life-long rare disease with other rare disease drugs and showed that, based on QALYs, a price of $4 million for this one-time treatment would be justified.

In fairness, Novartis executives did not say that this would indeed be the price of AVXS-101. It waited until after the drug was approved by regulators before announcing this. But Novartis clearly sent a signal to the world – especially payers – that we are heading into a new world of pricing not yet seen even for previously approved gene therapies.

It is pretty unusual for a company to discuss the price of a late-stage drug candidate at an analyst meeting intended to focus on R&D. Clearly, Novartis executives wanted to draw attention to the difficult issue of pricing new therapies that cure diseases as well as to begin to set expectations for AVXS-101. Perhaps the intent was to begin to engage a public debate as to the value of curing diseases. However, as unusual this is, effectively Novartis turned the tables on payers who have in the past begun to question the value of breakthrough drugs in advance of pricing discussions. We were now entering an era when the potential price of a drug gets debated publicly by the payers and drug makers before the therapy is even approved and the label is known.

On 24 May 2019, the FDA approved AVXS-101, to which Novartis gave the brand name Zolgensma™. Novartis also announced the price of this gene therapy – $2.1 million. While clearly less than the hypothetical $4 million price tag that Novartis had discussed six months earlier, it was still pretty shocking. Zolgensma™ became the world's most expensive therapy [4].

To ease concerns over this price, Novartis offered insurers the option to pay for the treatment in equal, annual installments over five years. The company also offered refunds if this gene therapy does not work.

The anticipated outrage over the Zolgensma™ price was somewhat muted thanks to an analysis from the usually conservative Institute for Clinical and Economic Review (ICER) [5]. Using a range of $100 000–150 000 per QALY, ICER gave a value-based price benchmark of between $1.1 and $1.9 million per treatment.

> "Zolgensma™ is dramatically transforming the lives of families affected by this devastating disease, and given the new efficacy data for the presymptomatic population, the price announced today falls within the upper bound of ICER's value-based price benchmark range," said Steven D. Pearson, MD, MSc, President of ICER.
>
> "Insurers were going to cover Zolgensma™ no matter the price, and Novartis has spoken publicly about considering prices that approached $5 million. It is a positive outcome for patients and the entire health system that Novartis instead chose to price Zolgensma™ at a level that more fairly aligns with the benefits for these children and their families" Source: "[5]/ICER.

Despite its record-breaking cost, Zolgensma™ was a cheaper treatment than the existing standard of care, Spinraza™. The latter is an antisense oligonucleotide marketed by Biogen. Spinraza™ must be given by intrathecal injection and is priced at $750 000 for the first year and then $375 000 for each year after. Spinraza™ is not a cure so in the first decade of treatment its total cost is $3.8 million. Zolgensma™ in comparison is a bargain at $2.1 million.

A third SMA treatment is now available, Evrysdi™ from Roche. It is an oral treatment that can be taken at home. It works by increasing production of a key protein, survival motor neuron (SMN), which is needed for nerves and muscles to function properly. Evrysdi™ is priced based on a patient's weight with a maximal price of $340 000 per year.

Interestingly, no single SMA treatment currently dominates the market. Figure 7.1 shows the breakdown of recent quarterly sales. It is surprising that a one-time treatment with Zolgensma™ has not dominated this market, especially as post-launch studies have shown that long-term safety and durability of efficacy are maintained [6].

Despite the examples of Kymriah™ and Zolgensma™, the pricing of gene therapies continues to be a major source of contention. The situation with bluebird bio's Zynteglo is a troubling case.

Beta-thalassemia is a blood disorder that reduces the production of hemoglobin, the key iron-containing protein essential for the transport of

Product	Patient Population	List Cost
Spinraza (Biogen/Ionis)	All SMA types, all ages	$750,000 in first year, then $375,000/yr.
Zolgensma (Novartis)	Type 1/2 children aged two or under	$2.1m one-off cost (spread over 5 yrs.)
Evrysdi (Roche/PTC)	All SMA types, all ages	Up to $340,000/yr.

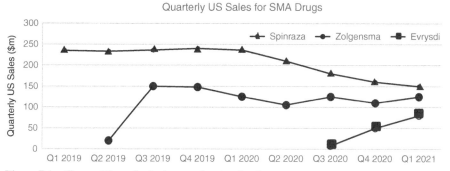

Figure 7.1   Competition of spinal muscular atrophy drugs.

oxygen to cells throughout the body. The diminished flow of oxygen causes weakness, fatigue, slow growth, bone abnormalities, and other serious complications. Affected individuals with severe symptoms require frequent blood transfusions in order to maintain a reasonable quality of life.

Bluebird bio received European regulatory approval for its beta-thalassemia gene therapy, Zynteglo, for patients 12 and over providing hope for these patients. A first of its kind treatment, Zynteglo allowed patients in clinical trials to go up to 56 months without the need for a blood transfusion – terrific news for these individuals.

However, the major attention that has been focused on Zynteglo has not been on the fact that it is a medical breakthrough but rather on its price. Given that it is a one-time treatment that might actually cure the disease, bluebird bio has set a price of $1.8 million making Zynteglo the second most costly treatment in history. Interestingly, despite this high price, some Wall Street analysts expressed disappointment as they felt that bluebird bio could have justified a much higher price.

Here is how bluebird bio rationalized Zynteglo's price [7]. A payment of $355 000 is made when a beta-thalassemia patient receives the infusion of the gene therapy. The remaining payments will be made to bluebird bio in four equal installments over the next four years providing there is no need for blood transfusions. The total due is $1.8 million – a sum bluebird describes as a 15% discount below an "intrinsic" value of $2.1 million, or 22 QALYs for successfully treated patients. In calculating the price of Zynteglo, the company specifically excluded the savings from the cost of treatment that Zynteglo replaces – which other companies have routinely included as part of the justification for such expensive therapies.

As would be expected, Zynteglo's price was not immediately embraced by the healthcare community. Dr. Peter Bach, director for the Center for Health Policy and Outcomes at Memorial Sloan Kettering Cancer Center told *STAT:*

> "Outcomes arrangements like these are an acknowledgement that the therapy is of uncertain durability, and while on their face they look like a good deal, they are actually a free option for the company to make a premium if the treatment works while not having to expend the funds to prove it works durably before selling it."

Bach is correct. But in fairness to bluebird bio, conducting long-term studies to demonstrate durability is not trivial and, if these were a preapproval condition, would have delayed getting this gene therapy to patients in need. However, there are some steps that bluebird bio could take to help address Bach's concern and that would help justify the cost of Zynteglo.

1. *No payment should be made until the patient is shown to have a medical benefit.* This is what Novartis has done with its CAR-T-based cure for childhood ALL and which should be emulated for all gene therapies. This would delay initial revenues for bluebird bio, but presumably it has sufficient confidence in its breakthrough that not getting paid until the gene therapy is proven to work is acceptable.

2. *bluebird bio should commit to re-treating patients at a 50% discount should the therapy wear off.* This addresses Bach's issue. Right now, we do not know if Zynteglo is a cure. If it is not, and if a "booster" might be needed at some point, a premium price should not be paid for additional treatments.

Bluebird has pledged to keep any price increases for Zynteglo below the Consumer Price Index. This is good news as the relative price for Zynteglo will remain constant in the coming years.

But even if bluebird bio were to make such commitments, it may not be enough to overcome price resistance. Although initially approved in Europe, the company announced that it would abandon this market due to what it called a hostile pricing environment [8].

Germany offered bluebird a $790 000 upfront payment for Zynteglo, which would go to $950 000 if the therapy worked after five years – far short of bluebird bio's $1.8 million asking price. It is likely that other European countries took a similar stance. It is not as if Zynteglo is an absolutely needed therapy. After all, beta-thalassemia is largely treatable with blood transfusions.

Trying to justify a very high price for a gene therapy is challenging when there are other treatment options.

One way to deal with the high cost of gene therapies is to use outcomes-based agreements (as was seen in Chapter 2 for the low-density lipoprotein cholesterol [LDL-c] lowering PCSK9 inhibitors). In such agreements, prices are agreed to by payers with the caveat that if the drug does not perform up to patients' and physicians' expectations, the company will refund part, if not all, of the cost of the drug.

Spark Therapeutics took such an approach when rolling out Luxturna®, a gene therapy that treats an inherited form of blindness. Spark set the list price for this one-time treatment at $850 000. Despite the price, Harvard Pilgrim Health Care became the first health plan in the nation to allow its members access to this important therapy, but they did so via an outcomes-based agreement. Here was the view of Harvard Pilgrim's Chief Medical Officer, Michael Sherman [9].

"We are thrilled to be able to offer our patients access to this groundbreaking genetic therapy. While this new treatment and other new therapies that are in the pipeline offer the promise of dramatic health improvements, their upfront costs are significant, which makes it imperative that we work together to find creative, value-based payment approaches that tie reimbursement to both short-term and long-term efficacy" Source: [9]/Harvard Pilgrim Health Care, Inc.

The agreement crafted by Harvard Pilgrim and Spark tied payments for the drug to measured improvement in patients at a 30- to 90-day interval and then again at the 30-month mark. If Luxturna® fails to perform, Harvard Pilgrim will receive a rebate from Spark. The short- and long-term measures are based on full-field light sensitivity threshold testing scores, with a baseline to be established for each eligible patient before administration of Luxturna®.

Clearly, a deal like this is an approach to pricing that can be tolerated by both parties. It is an approach that will likely be used more broadly as more gene therapies are brought to market.

In this discussion of gene therapies and their prices, the focus has really been on those patients who have adequate health insurance and who can afford any necessary co-payments that may be required to get the treatment. But, there are a lot of people in the world for whom gene therapies are not an option. This point was poignantly made in an editorial in *Nature Medicine* [10]. Sickle cell disease is a genetic blood disorder that affects predominantly people of African ancestries. There is a gene therapy in development that is showing great promise in curing this disease [11]. But, like other gene therapies, it is anticipated that the price tag for this one will be on the order of $1 million. That is going to be difficult for many of the 100 000 people in the United States who suffer from sickle cell. But such a staggering cost will be impossible for those in low-income countries and places where the long-term management of this disease is very challenging due to strained healthcare systems.

The editorial proposes that partnerships, for example among the Gates Foundation, the US NIH, and biopharma companies, could build collaborative international gene therapy research collaborations over the next decade to make such therapies more accessible. That is an interesting idea, but one that will not help those in dire need now. This serves as a reminder that scientific breakthroughs do not always promptly benefit the poor.

* * * * * * * * * * * * *

My guess is that my San Franciscan cab driver is not fully engaged with the gene therapy breakthroughs that are emerging. However, if he is aware that cures for certain diseases are available, he would not focus on the wonders of science but rather the outrageous prices that drug companies have instituted for these cures. He would likely be apoplectic at the thought of a single treatment priced at $2.1 million. Yet, as was described, this sort of price can be justified when it is truly life-saving.

We are now in an age when governments and payers are grappling with the value of a year of life. The United Kingdom has already provided its answer – $50 000 per year. It appears that the United States will eventually settle on a higher number. But, while these discussions are necessary, they seem harsh. Saving the life of a two-year old is priceless to the child's family. Restoring the sight to a blind person is invaluable. Extending by decades the life of a sickle cell patient is miraculous. Gene therapy is making all of this possible and giving life to people who will contribute to society in many, many ways. That is the true value of new gene therapies and they should be priced accordingly.

REFERENCES

1. Rosenbaum, L. (2017). Tragedy, perseverance, and chance – The story of CAR-T therapy. *New England Journal of Medicine* 377: 1313–1315. 5 October.
2. Hwang, T.J. and LaMattina, J.L. (2016). Measuring the Value of New Medications and Implications for Medicare's Proposed Part B Drug Payment Model. *JAMA Oncology* 2 (9): 1125–1126. 7 July.
3. Tong, A. (2021). Novartis' CAR-T fails to help a vulnerable group of NHL patients – squashing hope of moving up lines of treatment. *Endpoint News* (24 August).
4. Roland, D. (2019). At $2 million, new Novartis drug is priciest ever. *Wall Street Journal* (24 May).
5. Institute for Clinical and Economic Review (2019). Press Release, "ICER Comments on the FDA Approval of Zolgensma™ for the Treatment of Spinal Muscular Atrophy, 24 May, https://icer.org/news-insights/press-releases.
6. Johnson, V. (2021). Zolgensma™ shows safety and durability of efficacy in long-term follow-ups for SMA type 1. *Neurology Live* (16 March).
7. Feuerstein, A. (2019). Bluebird's gene therapy for a rare blood disease will cost $1.8 million. Cue the pricing debate. *STAT* (14 June).
8. Mast, J. (2021). Bluebird forced to halt trials after a cancer-like case pulls gene therapies from Europe following pricing backlash – shares slammed. *Endpoint* News (9 August).
9. Harvard Pilgrim Health Care Press Release (2018). Harvard Pilgrim is first health plan to directly negotiate outcomes–based agreement for groundbreaking gene therapy, 3 January.
10. Editorial Board (2021). Gene therapies should be for all, *Nature Medicine* Vol. 27, 1311 (1 August).
11. Kolata, G. (2021). Pioneering Gene Therapy Freed Her of Sickle Cell. Is a Cure at Hand? *New York Times* (14 September).

CHAPTER *8*

PROVING THE VALUE OF EXPENSIVE DRUGS: SHOULD WE PAY FOR DRUGS WHOSE ULTIMATE VALUE IS UNKNOWN?

IN GENERAL, the FDA has required companies to demonstrate both the safety and efficacy of new drugs before approving them. For example, as was discussed in Chapter 2, proof was needed to show that a promising new drug, torcetrapib, that both raised HDL-c (the so-called "good cholesterol") and lowered LDL-c ("bad cholesterol"), actually reduced heart attacks and strokes before the FDA would grant approval. Given that there were drugs already available to treat cardiovascular diseases, the FDA was not going to approve torcetrapib based solely on the effect it had on two major biomarkers of heart disease. That proved to be the correct decision as torcetrapib was less safe and effective than atorvastatin for heart patients.

Safety and efficacy are generally the norm for the FDA. However, it can choose to approve drugs based solely on biomarkers when the medical need is great and when there are no other options for patients. This is referred to as "accelerated approval," a program that was adopted by the FDA in 1992. The FDA took this stance with the Alzheimer's disease drug, Aduhelm™. However, the Aduhelm™ example is not unique. For a number of years, the FDA has been using accelerated approvals to make available cancer drugs based not on improving survival but rather on what is known as "response rate," that is, the percent of patients experiencing tumor shrinkage.

There are a variety of reasons why the FDA justifies granting an expedited approval for a drug based on response rate and not overall survival. First of all, some cancers grow slowly and it could take a number of years before it can be determined if the drug is actually improving survival. If it is found that a promising new drug is actually shrinking tumors, desperate patients and their oncologists would certainly want access to such a drug with the

Pharma and Profits: Balancing Innovation, Medicine, and Drug Prices, First Edition. John L. LaMattina.
© 2022 John Wiley & Sons, Inc. Published 2022 by John Wiley & Sons, Inc.

hope that it would, in fact, extend their lives. After all, these are patients that have a life-threatening disease and the FDA wants the chance to give these patients the opportunity to benefit from a potential breakthrough as soon as possible. But, proving that a drug actually improves overall survival is surprisingly challenging. This is due to the fact that many trials allow patients in the placebo group to move into the drug-treated group if early results suggest that the drug is working. This benefits the volunteer cancer patients in the trial, but it leads to inconclusive results in terms of demonstrating overall survival. Another challenge is that not all patients respond in the same way to drugs. It is not unusual for a subset of patients to thrive on the new cancer drug whereas others survive only months. Finally, it is unreasonable to expect cancer patients to wait years for the completion of confirmatory survival trials before being allowed access to it. They might not survive the wait.

What is wrong with the FDA trying to help terminally ill patients before there is definitive proof that an experimental cancer drug improves survival? They and their oncologists are certainly aware of the drug's potential albeit without conclusive proof. The answer is cost. Cancer drugs are generally priced in excess of $100 000 per year and critics argue that billions of health-care dollars are spent on drugs that are not proven to work [1]. Their concern is that the accelerated approval pathway is dependent on the completion of confirmatory trials that demonstrate long-term patient benefit as overall survival or improved quality of life. Unfortunately, of the 93 cancer drugs that received accelerated approval from 1992 to 2017, only 56 have completed confirmatory trials. Thus, oncologists are using drugs that they hope are effective but without conclusive proof that they do.

There are, in fact, examples where companies have pulled from the market drugs that had been initially cleared through accelerated approvals, but that later failed confirmatory trials [2]. For example, in 2016 Eli Lilly's Lartruvo had been approved for soft-tissue sarcoma – a rare cancer. The approval was based on a study of 133 patients who appeared to be benefitting from the drug. But the subsequent confirmatory study that involved 509 patients did not confirm the expected benefit. In 2019, Lilly asked the FDA to withdraw the drug from the market, which the agency did. This is how the system should work.

However, analyses have found that even when post-marketing studies demonstrate that drugs granted an accelerated approval have no clinically meaningful benefit compared to placebo, drugs retain FDA approval and remain on the market at prices comparable to those of the most expensive cancer drugs [3]. This should not be happening. These drugs are adding to cancer care costs, thereby burdening the overall healthcare system.

There are steps that can be taken to improve this situation. First, more rigor should be employed in both the design and reporting requirements for the confirmatory post-marketing studies. These trials need to include both overall survival and quality-of-life outcomes, which are crucial to cancer patients and their oncologists in weighing the benefits and risks of the drug. In addition, the FDA needs to set rigorous timelines for conducting these trials. Examples exist where these trials were not completed until after the patent for the drug has expired and the drug's sponsor has maximized its profits. That is not the behavior of an industry seeking to gain the public's trust. Finally, if a drug does not demonstrate the predicted benefits for survival and quality of life, it needs to be pulled from the market as soon as possible. This last point will be difficult to achieve if patients believe that they really are benefiting from the drug. In such cases, if the patient and oncologists are convinced the patient truly needs the drug, it should be provided at cost.

However, companies need to do more to gain the confidence of patients, physicians, and payers. Money-back guarantees could help. As was seen earlier, this was used by Amgen with its heart disease drug, Repatha®. In that case, if a patient has a heart attack while on Repatha®, the payer is eligible for a full rebate from Amgen for the drug's cost. There are stipulations. A patient must be on Repatha® for at least six months before being eligible for a full reimbursement. Also, each contract that Amgen signs specifies how long Repatha® is expected to protect the heart patient.

Novartis took a similar step with Kymriah®, the gene therapy for childhood acute lymphoblastic leukemia (ALL). This CAR-T-based therapy is priced at $475 000 per patient but no payment is made unless the patient responds to Kymriah® one month after treatment. If Kymriah® does not work, there is no charge.

Pfizer has now applied this concept to Xalkori®, which is used to treat non-small-cell lung cancer that has metastasized [4]. Pfizer is offering warranties for Xalkori® and will refund the entire cost to any patient and healthcare plan should the medicine fail to work after three months of treatment. For patients who discontinue use and document that the drug was ineffective, Pfizer will refund up to $19 144 for each month's supply, a total of $56 432 for three months, to cover out-of-pocket costs. This reflects the wholesale price for the drug. In addition, Pfizer will also refund the cost of Xalkori® paid by Medicare or the insurance company. What is attractive about this approach is that it provides refunds both to the patients and the payers. It also shows that the drug company is willing to share the risk with those taking its medicine.

The fact that Pfizer chose to do this with Xalkori® given Pfizer's stable of oncology drugs is interesting. Xalkori® has been on the market for over a decade and thus has a significant track record with respect to its efficacy and utility in the real world. It will be interesting to see if Pfizer expands this concept of warranties. If this program proves successful, it would put pressure on other biopharmaceutical companies to follow suit.

* * * * * * * * * * * * *

The FDA is often between a rock and a hard place when it comes to promising new drugs to treat diseases that can be fatal and for which no viable treatment exists. This is not only true for certain types of cancer but also for rare diseases. When any drug appears that offers the promise of alleviating such conditions, there are often patient advocacy groups that will intensely lobby the agency to allow broad access to what is still an experimental drug. No matter what the FDA decides in these cases, a large segment of the population will be unhappy. The view would be that the FDA is either lowering their standards or else the FDA is preventing patients in need from getting drugs that can save their lives. Unfortunately, there is no in-between view.

New drugs to treat cancer and rare diseases command high prices because of their therapeutic potential. But, drug companies need to share the risk. Offering some form of a money-back guarantee can make the high prices more palatable. Right now, the examples from Amgen, Novartis, and Pfizer are the exceptions. Making this a common practice will go a long way in restoring trust and confidence in the biopharmaceutical industry and its products.

REFERENCES

1. Frank, R.G.; Emanuel, E.J. (2021). Paying for cancer drugs that prove their benefit, *Journal of the American Medical Association*, doi: 10.1001/jama.2021.18308, 11 October.
2. Dooley Young, J. (2021). The FDA's accelerated approval process: When drugs are cleared for sale on limited evidence. *The Journalist's Resource*, https://journalistsresource.org/health/fda-accelerated-approval-drugs, 18 October.
3. Rupp, T.; Zuckerman, D. (2017). Quality of life, overall survival and costs of cancer drugs approved based on surrogate endpoints. *JAMA Internal Medicine*, 177, 276–277, 1 February.
4. Silverman, E. (2021). Pfizer will use a warranty to refund the cost of a lung cancer drug if it doesn't work. *STAT* (13 October).

CHAPTER 9

GENERIC DRUGS: BUILT-IN COST CONTROLS

I‌F YOU Google "The most hated man in America," one name consistently comes up – Martin Shkreli. In fact, *The Independent*, a British on-line newspaper opined that Shkreli was possibly the most hated man in the world [1]. Shkreli is not a mass murderer, nor the head of a major crime family. Rather, he is a former hedge fund manager and one-time CEO of a biotech company then known as Turing Pharmaceuticals. Back in 2015, Turing acquired Daraprim®, the brand name of pyrimethamine. Daraprim® is a drug that has been available for nearly 70 years and is used to treat a life-threatening disease, toxoplasmosis. This disease is caused by a common parasite, *Toxoplasma gondii*, and one can be infected by eating undercooked, contaminated meat or from contact with infected cat feces. Most people who are exposed to this parasite are asymptomatic, but it can cause serious complications in pregnancy, including stillbirth and miscarriage. In addition, people with compromised immune systems like those with AIDS or cancer are readily susceptible to this infection that, at the very least, can cause severe flu-like symptoms and lead to ocular damage, but in severe cases can be fatal.

After acquiring Daraprim®, Shkreli promptly raised the price of a tablet from $13.50 to $750. It is not as if Turing Pharmaceuticals had done anything new to improve the drug. It simply raised the price. The rationale used was that, as a life-saving drug, Daraprim® should be priced like other life-saving drugs such as novel cancer medications. That is a tough one to swallow for a drug that was just a few years earlier priced at $1/pill.

Yet, theoretically, this price increase should never have been possible. Daraprim® was originally discovered, developed, and manufactured by GlaxoSmithKline (GSK). The Daraprim® patent expired decades ago and the drug is now generic allowing any company to make it. However, it was such a small product that no one else bothered to compete. It simply was not worth it financially. In 2010, GSK sold the marketing rights to Daraprim® to CorePharma. This is not an unusual move for a big company. In order to

Pharma and Profits: Balancing Innovation, Medicine, and Drug Prices, First Edition. John L. LaMattina.
© 2022 John Wiley & Sons, Inc. Published 2022 by John Wiley & Sons, Inc.

optimize resources, companies will often divest minor products to small companies for whom the sales would be meaningful. According to the *New York Times* [2], sales of Daraprim® were less than $1 million in 2010 based on the price of about $1/pill. CorePharma raised the price to $13.50, which itself was surprising. But given the importance of the drug and the modest number of prescriptions per year (about 12 700), there was little complaint. That changed when the price was raised to $750/pill. When there was only one supplier of Daraprim®, Shkreli could control the price.

Finally, the FDA approved a generic version of Daraprim® in early 2020 produced by Cerovene, Inc [3]. One of the reasons for the delay was that the FDA generic approval process was backed up. But Turing (now known as Vyera Pharmaceuticals) also made things difficult for generic manufacturers by hindering them from getting samples of the brand medication needed to demonstrate bioequivalence. Cerovene did not announce its price, but it was expected to be 80–85% less than the Turing drug – still higher than pre-Shkreli times, but a distinct improvement. But the damage to the reputation of the whole biopharma industry had already been done. Industry critics continue to use Shkreli's behavior as typical of biopharma despite evidence to the contrary.

Shkreli's abuse of the system should really never occur, thanks to what Peter Kolchinsky calls "The Great American Drug Deal [4]." In his book of the same title, Kolchinsky persuasively argues both the importance of patents and their finite life.

> "The US government issues patents to allow inventors. . . . . . to profit from their invention for a defined period of time, typically 10 – 15 years after launching the drug. These temporary monopolies incentivize and fuel inventors and investors to develop new drugs. Companies that succeed in creating something new and beneficial can recoup their R&D costs, reward their investors, and plow some of those profits into future projects. This system keeps the biotechnology innovation engine churning and – importantly – prevents companies from milking cash-cow drugs indefinitely.
>
> Furthermore, this system creates a mountain of low-cost therapies that is the core of the Biotech Social Contract, which says, 'Yes, prices will be initially – and temporarily – high on branded drugs. But after the patent expires, society reaps the benefits of the cheap, effective generic versions for the rest of time.'"

A typical example occurred with Pfizer's cholesterol-lowering drug, Lipitor® (generic name, atorvastatin). Lipitor® had peak sales of $12.9 billion in 2007. Within a year of losing patent protection, Pfizer's Lipitor® sales dropped by 90%. In 2020, atorvastatin was the most prescribed drug in the United States with over 112 000 000 prescriptions written. A recent analysis of the 20 most prescribed medicines in the United States shows the dramatic impact that Amazon and GoodRx are having on the costs of generic drugs [5]. Amazon fills a 90-day prescription for a 20 mg atorvastatin tablet (the most common dose) for $4.20. That is less than $0.05/pill – truly a "Great American Drug Deal."

Kolchinsky makes a great comparison of generic drugs to hospital procedures, specifically hip replacements. Hip replacement surgery costs around $40 000, and that does not include the ancillary costs of rehabilitation, pain medications, antibiotics, etc. The first hip replacements were done in the 1940s and, while the artificial joints of today are much improved, the procedure is essentially the same now as it was then. There are about 400 000 hip replacements done in the United States annually, thus costing the healthcare system $16 billion. This cost will continue to rise and given the aging population and the ever-increasing annual hospital costs, the US hip replacement bill could easily reach $30 billion in the next decade. Hospital costs do not "go generic."

There are occasions when a branded drug still maintains significant share of the market despite the emergence of generic competition. That is what has happened with the famous little blue diamond pill, Viagra® [6]. Its generic form, sildenafil, was first available in December 2017. At that point, the cost of six tablets of 100 mg (the most prescribed dose) of Viagra® was $369. Generic sildenafil was initially $265 for six tablets. Since that time, however, the cost of the same six tablets of Viagra® has risen to $467 whereas that for sildenafil has plummeted to $4. One would think that, as a result of such steep competition, the Viagra® sales would have all but dried up by now. Yet, that has not been the case. The Pfizer spinout, Viatris, which now sells Viagra, will likely sell over $500 million of this erectile dysfunction drug in 2021, more than 40% of Pfizer's 2017 sales when the drug was still protected by its patent. How can this be?

For one thing, the price of branded Viagra® has been increased over the years. Any company can do this but usually little attention is paid. After all, as generic and much cheaper versions of the drug become available, few want the branded drug. Price normally drives the purchase of medicines. However, Viagra® is a very well-established brand. Furthermore, generic companies cannot sell sildenafil as a little blue diamond pill because its shape and color have been copyrighted by Pfizer. Instead, generic sildenafil tablets look different. Again, no one usually cares about the color and shape of a pill when they take it. I suspect that of the millions who take atorvastatin (generic Lipitor®), few remember what the original Lipitor® pill looked like. Also, in treating a disease like erectile dysfunction, there is the possibility of a "placebo effect" such that a generic pill might not be perceived as effective as Viagra® since it is not in the accepted little blue diamond. Given its aura, brand recognition, and higher price, Viagra® sales are still robust. However, as more pressure is put on healthcare costs, more and more sildenafil will likely be prescribed.

Help on providing low-cost generics is also coming from an unexpected source. Billionaire entrepreneur, Mark Cuban, plans to change the generic drug business paradigm with his newly announced venture: "The Mark Cuban Plus Drug Company [7]." For his first act, Cuban plans to produce albendazole, which, like Daraprim®, is an antiparasitic drug that retails for $222 per pill. Cuban's price will be $15. Cuban plans full transparency for the cost of developing and manufacturing generic drugs and will charge a flat 15% margin of the 100 drugs he plans to be soon producing. Cuban is taking an important step in helping to realize Kolchinsky's "Biotech Social Contract," by ensuring that important drugs that have finite proprietary lifetimes evolve into inexpensive medicines. Obviously, Cuban's enterprise will benefit patients and healthcare systems. But it will also benefit those research-based biopharmaceutical companies whose image should no longer be tarnished by the likes of Martin Shkreli.

To this point, the focus has been on generic versions of small-molecule drugs, the pills and tablets that people take. Manufacturing small-molecule drugs is very straightforward as the chemical processes used are familiar to all manufacturers and are highly reproducible. A generic manufacturer can make pills that are identical to those of the brand name maker thus ensuring that patients getting a generic drug will get the exact same medical benefits that are obtained for the brand name material. Furthermore, making small-molecule drugs is cheap, thereby enabling the 85% drop in price once generic competition unfolds.

However, over the past decades, another class of drugs has emerged – biologicals. These are large, complex molecules such as proteins and antibodies that are manufactured using living cells. These drugs are not administered orally. Rather, they are usually given in doctors' offices by infusion or injection. The manufacturing of biologicals is challenging since synthesizing drugs using cells cannot guarantee reproducibility. This is akin to wine making in that a fermentation process is used. The vintner's challenge is to ensure each new batch resembles the last. That holds true for biologicals as well.

Biologicals are far more expensive than pills due to higher research, development, and, especially, manufacturing costs. Biologicals make up about 2% of the total number of prescriptions written annually but a whopping 43% of overall drug spending [8]. Given the high cost of biologicals, one would expect that there would be a keen interest in having generic competition to these drugs once patents for the brand name expire. Such generic forms of biologicals are known as biosimilars, reflecting the fact that no two versions of a biological are identical.

However, the biosimilar market has been slow to take off. One reason is that the FDA was slow to approve biosimilars out of a concern that a biosimilar would not deliver the same efficacy and safety as a branded biological. The case of a Crohn's disease patient is revealing [9]. This 33-year-old woman had tried for years to find a medication to treat this inflammatory disease of the digestive tract when she finally hit upon Remicade® (generic name, infliximab). Remicade®, a TNF antibody, successfully controlled her disease. However, when the FDA approved a biosimilar to Remicade® known as Inflectra®, her insurance company had her switch to the newer and cheaper version. While the FDA's approval of Inflectra® stated that there were no meaningful clinical differences between the two, once this patient began treatment with Inflectra®, she suffered from a painful rash, fatigue, constipation, and nausea, all symptoms that did not occur with Remicade®. Cases like this make physicians and patients reluctant to switch from branded to biosimilar biologicals.

A second challenge in making biosimilars more prevalent is the patent thicket that companies have set up to protect their biologicals. AbbVie's Humira® (generic name, adalimumab), first launched in 2002, generates $20 billion in annual sales. This TNF antibody was first approved for the treatment of rheumatoid arthritis. However, subsequent approvals of Humira® for other inflammatory diseases such as Crohn's disease, ulcerative colitis, and psoriatic arthritis, as well as AbbVie's development of a novel auto-injector that made Humira® much easier to administer when compared to other TNF antibodies, have allowed AbbVie to extend the patent protection of Humira® to 2023. That is a tremendous boon to AbbVie, but also a tremendous cost to the healthcare system.

To date, 21 biosimilars have entered the US market [8]. On average, these are initially priced 30% less than the branded biologicals. However, over time it is expected that biosimilar prices will drop to roughly 50% of the branded product. That is less than the 85% drop seen when a pill becomes generic but, given the high prices of biological drugs, use of biosimilars can save the healthcare system billions of dollars. As more and more biologicals are being approved by the FDA each year, building a robust capacity to develop biosimilars will be important to keep drug costs under control.

* * * * * * * * * * * * *

The cost of drugs continues to garner headlines. However, methods already exist that could lower the burden of paying for medicines. Kolchinsky's "Biotech Social Contract" is based on allowing for competition in drug classes once a patent has expired. While this occurs with small-molecule drugs, improvements need to be made with biologicals. Perpetually high prices for these medicines are unsustainable. As Congress unfortunately debates ways to limit the profits of biopharmaceutical companies or to institute price controls, there are already ways to lower the nation's drug bill by the greater use of generics and biosimilars.

REFERENCES

1. Moore, J. (2017). Yes, he's the most hated man in America – possibly the world – but there's more to Martin Shkreli that meets the eye. *The Independent* (14 September).
2. Pollack, A. (2015). Drug Goes From $13.50 a Tablet to $750, Overnight. *New York Times* (20 September).
3. Chase, L. (2020). Daraprim®, the $750 Pill Finally Goes to Generic, https://www.goodrx.com.
4. Kolchinsky, P. (2020). *The Great American Drug Deal. A New Prescription for Innovative and Affordable Medicines*. Boston, MA: Evelexa Press.
5. Engelberg (2021). Outdated rule increases Medicare's costs to generic drugs by $26 billion a year. *STAT* (10 September).
6. Dotinga, R. (2021). Why Branded Viagra® Is Vastly More Expensive Than Its Generics. *MedPage Today* (15 June).
7. Silverman, E. (2021). Mark Cuban wades into the generic shark tank with a new company. *STAT* (13 January).
8. Goldman, D.P.; Philipson, D.J. (2021). Biosimilars Competition helps patients more than generic competition. *STAT* (8 October).
9. Andrews, M. (2021). Biosimilar drugs gain ground in U.S., but concerns linger about when patients can switch. *STAT* (21 September).

CHAPTER 10

ABOUT THOSE SOARING PHARMA PROFITS: ARE THEY DRIVING HEALTHCARE COSTS?

Iᴛ ɪs a critique often heard as pundits attack the cost of new drugs: the high price of drugs is fueling unseemly Big Pharma profits. Typical is the headline, "Drug prices rise as pharma profit soars [1]." While it is true that the cost of new drugs can be high (even if fully defensible), pharma profits are not greatly increasing as a result.

This dichotomy was forcibly argued by former Pfizer CEO, Ian Read, at a Forbes Healthcare Summit [2].

> "Is this industry obscenely profitable? There is no evidence of that. If you look at our return on investment, our return on capital, if you look at our P/E, if you look at anything inside this industry – looking at the Bloomberg indices – we are in the middle.
>
> So, I don't see an industry that you can say is profiteering. I see an industry that is taking its resources and investing into a high-risk business called 'innovation' and making modest returns on the capital at risk.
>
> So, I think the societal issue is how do you afford access to medicines that create great value, but require capital and risk to produce - the medicines that may represent 12 – 14 percent of the total costs and have automatic price adjustments in the form of loss of exclusivity? Source: [2]/Forbes Media LLC."

That is a pretty good speech but, in an era of fake news, how accurate are Read's comments? Actually, available data are pretty supportive. Return on equity (ROE), a measure of profitability, calculates the return on assets minus liabilities and, as Read stated, the biopharmaceutical industry does not stand out. Data from Stern NYU [3] show Pharmaceuticals have an ROE of 12% (adjusted for R&D). This is comparable with Healthcare Support Services at 16.6% but less than Hospitals/Healthcare facilities (70.6%), Retail/Automotive (36.2%), Computers/Peripherals (29.6%), Beverages/Soft

Pharma and Profits: Balancing Innovation, Medicine, and Drug Prices, First Edition. John L. LaMattina.
© 2022 John Wiley & Sons, Inc. Published 2022 by John Wiley & Sons, Inc.

(27.8%), and Semiconductor Equipment (21.2%). The Biotechnology sector's ROE is a mere 1.3%, reflecting the fact that many of these companies have yet to market a drug. Based on ROE, the biopharmaceutical industry is essentially in the middle when compared to other industries.

Another measure, Internal Rate of Return (IRR), is even more telling. IRR calculates the sales/cash flows resulting from R&D investments, ties R&D and the returns it generates together, and is a more appropriate metric for biopharma productivity. Deloitte reports that the IRR for biopharma R&D was steadily falling from 10.1% in 2010 to 1.6% in 2019 [4]. The trend reversed a bit in 2020 growing 2.5%, but still off where it was a decade ago, Again, the biopharmaceutical industry does not stand out as overly prosperous.

Read's final point is also on the money. Data from the National Health Expenditure table issued by the Centers for Medicare and Medicaid Services show that in this century, the percentage of money spent on prescription drugs has consistently been in the 12–14% range, as shown in Figure 10.1 [5].

There is no doubt that healthcare is very costly in the United States. However, drugs make up a relatively small percentage of the nation's healthcare bill. If politicians were magically able to eliminate all drug spending, healthcare costs would still rise. In seven years, that bill would reach today's level, as the non-drug components of healthcare spending continue to grow rapidly.

Total drug costs are, however, moderating. This is shown in data recently released by the Altarum Institute, a nonprofit research and consulting company that serves government health insurers, health foundations, and nonprofits that focus on health and healthcare. From March 2020 to March 2021, the HealthCare Price Index – a composite of all healthcare costs – rose by 2.5%. Yet, over this same timeframe, prescription drug costs *decreased*

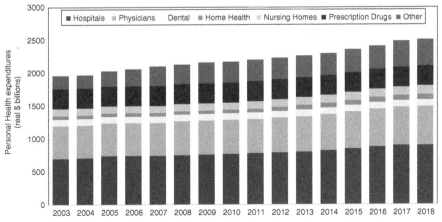

Figure 10.1   Healthcare spending 2003–2018. Source: Modified from [5].

by 2.3%! The main drivers for the increase in healthcare spending were hospital costs (4.8% increase) and physician and clinical services (up by 2.9%) [6].

Altarum's data show that 2020/2021 was not an aberration. In March 2019, the national health expenditure for drugs dropped by 0.4% while March 2020 showed an increase of 1.5%. Thus, on balance, drug spending has essentially been flat for at least three years. In contrast, over 2019 and 2020, hospital care spending grew by 1.3 and 2.5%; physician and clinical services by 0.5 and 1.2%, respectively.

The discrepancy between spending on prescription drugs and hospital care is even more pronounced when comparisons are made with other countries. Dr. Elias Zerhouni, former Director of the NIH and president of Sanofi R&D, showed some surprising data at the 9th Annual International Symposium of Drug Delivery Systems at Stanford in his talk entitled "Drug Discovery and Bio-innovation: Opportunities and Challenges [7]." Concerned about unsustainable healthcare costs, Zerhouni contrasted healthcare costs between the United States and France. Not surprisingly, France spends less on drugs ($750 per person) than the United States ($1450). But this twofold difference pales in comparison to hospital costs. A colonoscopy in France costs $400 compared to $2000 in the United States. An angioplasty in France is $7000; $25 000 in the United States. A simple MRI costs $1200 in France, $4300 in the United States.

As was pointed out in Chapter 9, there is a built-in cost control mechanism for all brand name drugs – patent expiration. After a drug's patent expires, its price in the United States can drop by as much as 90% within a year. There are no such cost containments for colonoscopies, angioplasties, or MRIs. Procedures like hip replacements do not "go generic." Rather, they will grow in expense by about 5% annually.

* * * * * * * * * * * * * *

When people hear about the high cost of new medicines such as a $1000/pill for a drug to treat hepatitis C or $2.1 million for a gene therapy to cure a childhood leukemia, the natural tendency is to think that drug companies are wildly profitable. After all, given such high prices, huge profits must surely follow. But, as was said at the outset, investments in the biopharmaceutical industry come with great risk. This is an industry where fewer than 10% of therapies that enter clinical trials actually succeed. The business of biopharmaceutical innovation is expensive and time-consuming.

According to data from NDP Analytics (research supported by the Pharmaceutical Research and Manufacturers Association – PhRMA), the biopharmaceutical industry has relatively few profitable firms [8]. Only 14% of publicly traded biopharmaceutical companies showed positive net income in 2017 compared to 65% of companies in all other industries. That same year, the top 50% of 728 publicly traded US biopharmaceutical companies had a weighted average profit margin of 7.5%, while the bottom 50% had a loss margin of minus 11 000%. This is a tough business.

It is ironic that while criticism of drug prices is growing, spending on drugs has been moderating. Perhaps this is due to public pressure, and perhaps that moderation is a good thing. However, the bigger components of healthcare spending continue to rise. The second irony is that drugs (relatively inexpensive) can actually help to prevent hospitalizations and complications of diseases (much more expensive). One can argue that we should be doing everything to *maximize* the use of these medications. But the target of cost containment pressures continues to be on drug costs. Sadly, legislation is proposed that will essentially drive the United States down the road toward price controls. That will be addressed next.

REFERENCES

1. Kacik, A. (2017). Drug prices rise as pharma profit soars. *Modern Healthcare* (28 December).
2. LaMattina, J.L. (2018). About those soaring pharma profits. forbes.com (23 January).
3. NYU Stern (2021). https://www.stern.nyu,edu/~adamodar/pc/datasets/roe.xls (January 2021).
4. Deloitte (2021). The Seeds of Change – Measuring the return from pharmaceutical innovation 2020. https://www2.deloitte.com<Deloitte>documents.
5. Baicker, K. and Chandra, A. (2020). Do We Spend Too Much on Health Care. *New England Journal of Medicine* 383 (7): 605–609. (13 August).
6. Rhyan, C. (2021). Perspective: Are Rising Health Care Prices Another Covid-19 Side Effect. https://altarum.org/news/are-rising-health-care-prices-another-covid-19-side-effect (16 April).
7. LaMattina, J.L. (2021). Hospital costs, not drug prices are out of control. forbes.com (26 April).
8. Pham, N.D.; Donovan, M.; Pierce, B. (2019). A comparison of profitability measurement between R&D intensive and non-R&D intensive industries. https://ssrn.com/abstract=3352831.

CHAPTER *11*

SCHEMES TO LOWER DRUG PRICES: THE IMPACT OF REDUCED RESOURCES ON R&D

Gɪᴠᴇɴ ᴛʜᴇ concerns about the high price of medicines, a number of proposals have been made that seek to lower these costs. One idea is to import drugs from Canada. After all, that is effectively what Senator Bernie Sanders did when he took a group of Minnesotans across the border to get their insulin analogs (Chapter 5). While president, Donald Trump supported a Florida proposal to import prescription drugs from Canada.

However, this route involves considerable challenges. The major one, which was correctly advanced by the Pharmaceutical Research and Manufacturers of America (PhRMA), is that importing drugs is dangerous, even if the importation is through Canada. PhRMA noted that in 2018 the FDA caught 3600 websites selling illicit and counterfeit medicines. But this is not just the view of PhRMA. An op-ed piece in the *New York Times* by Katherine Eban [1] questioned the safety and viability of drugs made overseas and shipped to the United States. Eban showed the fraudulent quality-control data coming from overseas manufacturing plants calling into question the integrity of these drugs. In addition, she provided examples of insect and bird infestations in these plants, as well as toxic impurities in drugs being sent to US patients. These drugs clearly are not of the quality we expect for our medications. While some may hope that Canada would be a safe source for drugs, PhRMA has pointed out that Canada has stated that it is not able to certify that medicines shipped through its borders to the United States would be safe. Guaranteeing the safety of imported drugs is a big challenge.

Even vaccines and medicines for Covid-19 are being impacted by counterfeiters [2]. Substandard and falsified products are emerging, which can undermine the efforts to control the pandemic. Ensuring that Americans get only medicines that have undergone strict quality-control measures is critical

Pharma and Profits: Balancing Innovation, Medicine, and Drug Prices, First Edition. John L. LaMattina.
© 2022 John Wiley & Sons, Inc. Published 2022 by John Wiley & Sons, Inc.

in treating all diseases, not just Covid-19. Because the supply chains of ethical companies are bypassed, importing drugs through Canada jeopardizes patients.

Pharma's critics have a response. They suggest that Canada (and its regulators) acts as a supplier. Unfortunately, that will never happen. Drug manufacturers manage shipments and supplies for each country. Should the state of Florida go ahead and allow the importation of drugs from Canada, and should other states follow suit, these actions would quickly empty the shelves of Canadian pharmacies. Manufacturers are not going to suddenly ship more drugs to Canada so that they can be reshipped to the United States at lower prices. They will limit supplies to Canada to the levels that were sold before US re-importation began. The Canadian government, faced with drug shortages for its own population, would then block the export of those medicines. Clearly, importing drugs from Canada (population: 38 million) will not solve drug pricing issues in the United States (population: 334 million).

Another cost-control proposal centers around "international reference pricing" (IRP). Enactment of such a scheme would benchmark US drug prices to those prices in foreign countries. In effect, the US prices would be set by other nations.

Alex M. Azar II, former Secretary of Health and Human Services (HHS) under Donald Trump, derived a version of IRP to limit prices [3] by proposing changes that would enable the government to reduce payments for the 27 highest-cost physician-administered drugs – drugs that are part of the Medicare Part B program. His rationale was simple:

> "In Medicare Part B today, the government gets the bill, and we just blindly pay it, plus six percent for the provider who administers it. There is no negotiation, and the payment mechanism actually encourages prescribing more expensive drugs."

HHS analyzed the prices that Medicare pays for these 27 drugs and found that, on average, they were 180% of what other wealthy countries pay. Azar's proposal would change this by requiring manufacturers to provide discounts over a five-year phase-in period so that Medicare would pay 126% of what other countries pay.

Needless to say, the biopharmaceutical industry had a number of issues with Azar's proposal. PhRMA argues that US prices should not be determined by countries with wholly different market conditions. For example, Western Europe has single-payer health systems – monopsonies that set modest prices. Less wealthy countries like Greece can insist on even lower prices. Interestingly, the PhRMA concerns have been echoed by both the *Wall Street Journal* and the *New York Times* [4, 5].

Azar grossly underestimated what the impact of his proposal would have on the biopharmaceutical industry.

> "Our model will save $17 billion in Medicare drug spending over the next five years. That's $3.4 billion a year. The pharmaceutical industry reports they spend an average of 21 percent of revenue for R&D. At most, this model could pull around $700 million out of their annual R&D budgets, which they boast are more than $70 billion a year. These savings, while substantial for American patients and taxpayers, cannot possibly pull out more than one percent of R&D."

Thus, Azar claims that HHS can reduce spending on Medicare Part B drugs with little impact on biopharmaceutical R&D investment. Duane Schulthess, Managing Director at Vital Transformation, questioned Azar's claim on this. As he delved into specific compounds on Azar's list of 27, he found some startling data. One striking case was with Biogen's Multiple Sclerosis drug, Tysabri.

> "According to Biogen's 2017 annual report, sales of Tysabri represented $2 billion of their global revenue of $12 billion. In other words, roughly 17 percent of Biogen's total revenue was derived from this one product in fiscal year 2017. $1.12 billion of Biogen's revenue from Tysabri came from the U.S.; the rest of the world was responsible for roughly $860 million. While only $306 million of their U.S. sales were accountable to Medicare Part B, there is little doubt that a forced reference price would radically impact all U.S. prices, as any insurer would negotiate based upon the price paid by Medicare Part B.
>
> In 2017, Biogen spent $2.26 billion on R&D. A cut of $630 million in U.S. revenue due to reference pricing equates to a whopping hit of 30 percent of Biogen's total R&D budget, not one percent."

How does Schulthess come up with a $630 million revenue decline in Tysabri sales under IRP? The US price of Tysabri is 2.9 times higher than the average benchmark in Europe. Thus, if one "normalized" the $1.12 billion in US sales to the European levels, Tysabri sales would be lowered to $386 million in the United States. However, Azar has said that Medicare would allow a 26% markup over the European prices. Thus, assuming that all US payers would benefit from the price set by HHS for Medicare Part B, Tysabri US sales would drop from $1.12 billion to $486 million, a loss of over $630 million. Given that Biogen invests roughly 20% of top line revenues into R&D, this drop in Tysabri sales would amount to $126 million less for R&D. That is a hit of 5.4% of the R&D budget, not 1% (but not 30% either).

Such a loss in research funding for Biogen is not trivial and would have severe consequences. As a former R&D president, I know that even a 5% funding hit leads to research programs being cut and scientists being let go.

But the loss of $630 million in sales would reverberate not just in R&D but throughout the whole company with job eliminations across the board. Wall Street would also react negatively limiting Biogen's growth prospects.

Unfortunately, this threat is not unique to Biogen. There are companies like Amgen and Roche with multiple entries on Azar's list. Once again, we see unintended ironies. Tying Medicare Part B drug pricing to IRP will disproportionately hurt companies that offer more innovative therapies for the most challenging diseases. Should a proposal like Azar's come to pass, there might be unintended consequences. Biopharmaceutical companies could shift R&D resources away from diseases managed by Medicare Part B physician-administered drugs – areas where we are still in dire need of new therapies. This would be tragic.

Faced with drug-price criticism, the biopharmaceuticals argue that any such controls will hinder the search for new, life-saving medications. This claim is usually rebutted by critics who claim exaggeration and bluster. They liken the industry to Aesop's boy who cried wolf. They tire of biopharma's "innovation card."

The following account demonstrates what happens when R&D budgets are cut significantly. In late 2006, Pfizer learned that a potential blockbuster cardiovascular drug, a drug that essentially remodeled patients' lipoprotein profiles, had no effect in prolonging the lives of people with heart disease. This drug had been expected to be a key driver of future revenues and propel Pfizer's growth over the coming decade. These clinical results were devastating. The chief executive decided the company had to be "right sized" based on decreased future projected revenues. In other words, budgets across the entire company had to be cut – including that of R&D.

At the time, I was in charge of global R&D from the initial discovery stage through regulatory approval. My budget was just under $5 billion. (Pfizer's overall corporate R&D budget was close to $8 billion, but this number included the R&D budgets of other divisions, the costs of post-marketing clinical trials, research expenses for outside investments, etc.) Going into 2007, R&D was seeking a 10% increase. That is a pretty hefty number based on a $5 billion base. However, this increase was needed to maintain the ongoing programs, many of which were in key – and expensive – clinical trials. However, given the new reality, budget increases were denied. Instead, R&D was told to cut its budget by 10%. So, a vision of a $5.5 billion to run a global organization of 13 000 people and a portfolio of over 100 compounds in various stages of development, was shattered by a $1 billion shortfall.

Maybe our critics cried: "Big deal! Large companies are inefficient and unwieldy. Certainly, there must be plenty of fat that can be cut." The reality was very different. In 2000 and then again in 2004, Pfizer made major acquisitions, first Warner-Lambert Parke-Davis and then Pharmacia. In each case, to justify to shareholders the wisdom of these mergers, "synergies" had to be found. The organization had to remove overlap and cut programs after vigorous prioritization processes. As a result, a lot of cost cutting had already taken place. Sure, there are always cuts that can be made. But simple trimming was not going to provide $1 billion in savings.

It became pretty apparent that dropping programs would not be enough. The new budget demanded that we eliminate "bricks and mortar" – research sites and the jobs of scientists who worked in these labs. Smaller sites were the obvious choices including two in France and one in Japan. But this did not provide nearly enough savings. Eventually, we decided to close the legacy Parke-Davis labs in Ann Arbor, Michigan, and eliminate that site's 2200 jobs. Coming at a time when the auto industry was experiencing its own economic issues in Michigan, this closure generated serious concerns across that state.

Furthermore, estimates are that for each job that exists at a major research site like Ann Arbor, three other jobs are created in the community. Thus, the closing of the Ann Arbor labs had a big impact on other sectors of the local community.

But the most difficult part of this downsizing was the loss of thousands of excellent scientists. These great colleagues were leaving the company. It is incredibly hard to witness the lives of our friends disrupted, their families uprooted.

Changes to the Pfizer discovery research portfolio were also made. Major discovery therapeutic areas were shuttered: metabolic diseases (cardiovascular, diabetes); CNS (depression, anxiety, neurological disorders); infectious diseases (antibiotics, antifungals, antivirals); and allergy and respiratory diseases. These areas had produced some of the world's most important drugs such as Lipitor®, Zoloft®, Norvasc®, Diflucan®, Zithromax®, and Lyrica®. With a smaller R&D budget, Pfizer could not afford to operate across multiple therapeutic categories.

At the peak of its R&D investments in their labs in Groton and New London, Connecticut, Pfizer employees were generous charitable givers. During that time, Pfizer made up 5% of the jobs in New London County, 10% of the salaries, and 55% of the local United Way donations. With the downsizing of these labs, the United Way saw a dramatic decrease in its fundraising impacting the programs it supported.

Sometimes, people assume pharma companies are insanely rich companies and can easily absorb revenue cuts. Few, if any, recognize the ripple effect suffered by many who support these companies and who live in the communities that house these research labs.

Nevertheless, plans continue to find ways to drive down drug prices. The one generating a lot of momentum currently is to enact a law to enable

Medicare to negotiate drug prices with the biopharmaceutical industry. In fact, a Kaiser Family Foundation (KFF) poll found that 83% of the public supports such a measure [6]. Unfortunately, most adults do not believe high drug prices are needed for drug companies to invest in new research. Instead, people believe that, even if the US drug prices were lower, drug companies would still make enough money to discover and develop new drugs.

There is a major misconception that drug companies could be much better at picking pipeline "winners" so that "losers" could be set aside. Unfortunately, predicting the importance of a new experimental medicine is next to impossible. It is like looking at a group of first graders and predicting who will make the most important contributions to society in their lifetime. Sometimes exploratory drugs thought to be major breakthroughs fizzle out. Other times, as happened with Lipitor®, a drug with modest expectations blossoms into a clinical and financial super star.

Industry critics imply that drugs that are not first-in-class – so-called "me-too drugs" – offer little value. But that thinking is wrong. The first entry in a new class of drugs is rarely the best. In the case of the cholesterol-lowering statins, Lipitor® and Crestor® were the fourth and fifth statins approved by the FDA. Yet they turned out to be superior to the earlier entrants and have had major impacts on the treatment of heart disease.

And, it turns out, having multiple drugs in a class is actually important for controlling drug prices. When Sovaldi®, the cure for hepatitis C, was introduced in 2014, a 12-week course of treatment had a list price of $84 000 – $1000 a pill. However, with the advent of other hepatitis C drugs, like AbbVie's Mavyret®, the list price for newer hepatitis C drugs dropped to $24 000 for an 8-week course – a little more than $400 a pill. Competition is critical for lowering prices.

One might claim that the ultimate me-too drug is Spikevax, Moderna's Covid-19 vaccine. After all, it was approved after Comirnaty, Pfizer's vaccine (albeit by a week). Both mRNA vaccines behave similarly with respect to safety and efficacy. Of course, in a pandemic, we want as many vaccines as possible including so-called me-too vaccines. But this example demonstrates that multiple companies are often working on the same program, work that starts years before a product crosses the finish line. The successful ones can reach regulatory approval within 12–18 months of each other. It is not as if companies intentionally set out to create me-too drugs. Yet having multiple options for a new class of drugs is just as important for patients – no one drug is perfectly suited for an entire population. Further, choice in a drug category is a boon to payers.

Legislation being considered by Congress will have an impact on drug prices. Three general changes are being proposed, all designed to help the consumer. One part of the law would control the rate of price increases in both Medicare Part B (physician-administered drugs, usually biologicals) and Medicare Part D (drugs that are self-administered). Manufacturers of drugs sold to either program that raise the relevant prices of their drugs more rapidly than inflation would have to pay the above inflation amount back to the government [7]. Another part of the law creates an out-of-pocket cap for Medicare Part D beneficiaries at $2000/year as well as a $35 monthly cap on insulin expenses. Currently, no such caps exist.

But the third pillar of the law is the most concerning for drug makers. Beginning in 2025, the United States will begin negotiating the price of the 10 most expensive drugs, a number that will rise to 20 drugs by 2028. The caveat is that drugs subject to price negotiation must be nine years past their exclusivity period (patent life) and 12 years for biologics. This is a modest start. But the obvious concern is that, once this door is opened, government price restrictions could soon be applied to all medicines.

There is no doubt that allowing Medicare to negotiate drug prices will have a major impact on the productivity of biopharmaceutical R&D. The question is how big of an impact will this be. Various analyses have been done to calculate this. For example, one by the Congressional Budget Office (CBO) suggested that in the first decade of the enactment of such legislation, there would be 2 fewer drugs produced in the first decade, 23 fewer drugs in the second decade, and 34 fewer in the third decade [8]. However, another analysis, this one from the University of Chicago [9], paints a much bleaker picture. Using literature data, these authors demonstrate that enactment of broad Medicare negotiation legislation will lead to a 29–60% reduction in R&D from 2021 to 2039, which would translate into 167–342 fewer new drug approvals during that period. Even if you take the most conservative numbers from the professors at the University of Chicago, the loss of 167 new drugs would be tragic, especially coming at a time when we are learning more and more about the causes of the most difficult diseases to treat. The impact on patients would be devastating.

To put the importance of biopharmaceutical R&D into perspective, it is estimated that the industry invests over $160 billion each year. For comparison, in the last fiscal year, the National Institutes of Health budget was $41 billion. The CBO estimates that the biopharmaceutical industry invests 25% of its revenue into R&D, the highest percentage by any industry [10]. Given the importance of biopharmaceutical R&D on the world's health, one could argue that this industry is the jewel of the United States. Why would we want to discard this treasure?

What would likely be the outcome of cuts to revenues of biopharmaceutical companies? To a certain extent, the Pfizer experience discussed earlier serves as a model. First of all, investing 25% of revenues into R&D as now happens would not be sustainable and this would drop perhaps to historical levels of 15%. There will be the double whammy of less revenue due to price cuts coupled with a lower investment rate in R&D. The industry's overall investment of $160 billion would be hit. Companies would have to keep their focus on drugs in the late-stage research pipeline as those offer the most potential for new revenues in the not too distant future. Thus, the R&D cuts would have to come from the early stages of R&D. The industry would, in effect, be eating its seed corn and jeopardizing its future.

The threat to established companies is obvious. Perhaps more worrying is the potential destruction of the biotech sector, dominated by small start-up companies, many operating on borrowed cash. The legislative proposals will certainly alter the favorable views that venture capitalists (VCs) currently have for biopharmaceuticals. The *Wall Street Journal* accurately summed up this potentially dire situation as follows [4].

> "This tax on innovation would damage big pharma companies like Pfizer and Merck, but it would be a death blow to small biotech firms where many breakthroughs originate. Biotech startups patented nearly two-thirds of new drugs in 2018, and in their early stages they rely almost entirely on venture capital. Later they often cooperate with or license their discoveries to large drug makers. Only about 15 percent of drugs and vaccines that enter clinical trials are approved. The rate for oncology therapies is three percent. If Democrats slash the return on drug investment, startups won't be able to raise money. This would be a tragedy as we enter an age of great biotech discovery Source: [4]/ Dow Jones & Company, Inc."

Unpredictable government-dictated prices would give VCs pause in supporting biotech start-up ventures and instead they would shift their investments toward technology and consumer goods. The robust early ecosystem that we now enjoy, which helps to explore early medical hypotheses, would be damaged – perhaps irrevocably.

Another impact of decreased revenues would be that bigger companies would shift their research priorities. With fewer dollars to invest, companies would shun therapeutic areas where long and expensive clinical trials are needed to prove the long-term benefits of a new therapy. With a limited R&D budget, it is hard to justify beginning a new program to treat heart disease (still the world's leading killer) when a billion-dollar cardiovascular phase 3 trial will be needed for FDA approval, followed by pricing negotiations with payers. Instead, companies will focus their resources in therapeutic areas where clinical trials would be shorter and cheaper, and where the drug would command a high price. Therapeutic areas like cancer, rare diseases, and vaccines would thrive, but cardiovascular diseases, diabetes, obesity would take a hit. In addition, companies might shift R&D resources into areas where patients would be willing to pay for the drug themselves such as medical aesthetics and sexual function.

As revenues decrease for the biopharmaceutical industry, there will be downsizing as happened at the Pfizer Ann Arbor labs. Envision that experience occurring at research sites across the country. The impact will be felt not just on the company employees but throughout the community. When Pfizer downsized back in 2007, many of the scientists were able to get research jobs at other biopharmaceutical companies. However, if the whole industry were to undergo a similar situation, the options for scientists would be limited. Many will likely need to change careers. A generation of drug discoverers will be lost.

To be clear, all members of the biopharmaceutical ecosystem care about high drug prices. If major medical breakthroughs are made that lead to important new medications, they are useless if they are unaffordable. If patients cannot get access to these breakthroughs, then we have failed in our life's work.

Furthermore, there is no doubt that Americans pay more for their drugs than patients in other parts of the world. As shown in an analysis by the RAND Corporation [11], drug prices in the United States exceed those of most countries.

"The findings indicate that 2018 drug prices in the United States were substantially higher than those in each of 32 comparison countries when considering all drugs together. Compared with all comparisons countries combined, U.S. prices were 256 percent higher than those in other countries."

It should be noted that the RAND analysis is flawed as they use *list* prices in their calculations and not the *net* prices paid for these drugs. It must be stressed that most patients do not pay list prices due to rebate to commercial payers, patient assistance programs, etc. In fairness, RAND was not trying to be misleading in its analysis. The problem is that net prices are not readily available as these data are held in confidence. Both manufacturers and payers keep these figures secret as this is a key component of their negotiations. This does not happen in the United Kingdom, France, and all the other countries that have single-payer systems. This is not to say that Americans do not pay more for drugs; we do.

The United States is paying for the innovation to produce new medicines that benefit the rest of the world. The United States also supports the world in other areas including defense. But, that fact does not resonate with Americans when they continue to dig deeper into their own pockets to pay for their drugs. This is not lost on Big Pharma CEOs who are looking to ease these cost burdens on patients. Pfizer CEO, Dr. Albert Bourla, recently addressed this issue in the following way [12]. He pointed out that Pfizer *net* prices *decreased* by 5% in the first six months of 2021, a trend that has been happening for a few years. He went on to say:

> "I'm sure the patients taking our medicines didn't experience this minus five percent. They experienced two, three, five, 10, 20 percent increases in how much they have to pay for the cost of the medicine."

Given this situation, it is no surprise that, in 2019, Americans paid a record $67 billion out of pocket for prescriptions filled in retail pharmacies. Bourla commented:

> "Americans are paying for their medicines like they don't have insurance, although they do have insurance. And this needs to change."

Rather than focusing on controlling drug prices, Congress needs to focus on the high out-of-pocket costs being inflicted by insurers. Legislation that provides for out-of-pocket monthly caps is a good step. Also needed is a requirement that payers pass on annual cost savings they accrue as a result of price negotiations with the biopharmaceutical industry. What good is having healthcare insurance if the insurance does not pay for the medicines that you need to survive?

Congress can also do more in speeding access to generic drugs. The aforementioned RAND Corporation analysis found that this is one area where Americans are paying less than in the rest of the world. Generic drugs, which make up more than 85% of annual US prescriptions, are priced on average in the United States at 84% of those same drugs in other countries. For the "Great American Drug Deal" (outlined in Chapter 9) to be realized, drugs have to go generic without undue delay. This is the core of the "biotech social contract" and needs to be addressed.

Finally, biopharma should accept the proposed cap on annual drug-price increases. Industry should limit price increases for drugs purchased by Medicare to a measure of consumer inflation. This would be a big help in helping the industry show the public that it cares about affordable medicines and is willing to take steps to help address the situation.

There are a number of schemes being put forward to limit the cost of life-saving medications. These are all being put forward with the assumptions that the biopharmaceutical industry can easily adjust to a decrease in revenues. Unfortunately, this, at best, is a gross oversimplification and, at worst, will have major consequences for an industry that deserves to be cherished. Stories about applying cost-cutting measures for drugs enter with the bias that:

1. This is a very wealthy industry.
2. Paring its revenues will be painless.

3. The impact on the industry's productivity will not be significant.

4. The industry simply needs to work on the "right" projects and not the losers.

These are all incredibly ill-informed and misguided assumptions.

The knock-on effects of any decrease in biopharmaceutical revenues are significant: jobs will be eliminated, research opportunities will be lost, fewer new medicines will be developed, and communities that host research centers will be affected. Depending on the response of VCs, the US biotech industry, currently the envy of the world, will be devastated. Is this what we want at a time when the opportunities to discover cures for diseases have never been higher?

Dr. Howard Dean is the former Governor of Vermont, chair of the Democratic National Committee from 2005 to 2009, and a candidate for the Democratic nomination for President in the 2004 election. A few years ago, he shared his views on drug-price controls in a letter to the *New York Times* [13].

"The American drug industry is by far the most successful and innovative in the world in addition to being the most expensive because we are the only country that pays the true research and development costs, not only for Americans, but for the rest of the world as well...Limiting profits may sound attractive but it will also be ineffective. The more honest discussion is about what we as a society are willing to pay to improve or extend life. And the answer is a lot, which is why reform is always talked about but never accomplished...

...But schemes to launch a federal attack on one of the last growing innovative industries in America are in the long run counterproductive for both job creation and, more important, for the health of human beings around the world. By all means let us try to reduce the cost of drugs. But over the years, advances in drug efficacy and scope have saved us far more in hospital costs than we have spent on drugs Source: [13]/The New York Times Company."

Amen.

REFERENCES

1. Eban, K. (2019). Americans Need Generic Drugs But Can They Trust Them. *New York Times* (11 May).
2. Zaman, M.H.; Sundaram, R.; Gabriel, W. (2021). Fake, substandard vaccines and medicines spell trouble for controlling Covid-19. *STAT* (25 October).
3. LaMattina, J.L. (2019). Azar's plan to tie U.S. drug prices to foreign ones will impose investments in R&D. *STAT* (21 June).
4. Editorial Board (2018). Why are drugs cheaper in Europe? *Wall Street Journal* (28 October).

5. Pear, R (2018). Trump officials say drug prices are inflated. So are some of their claims on a solution. *New York Times* (16 December).

6. Kirzinger, A., Kearney, A., Stokes, M. et al. (2021). Public Weighs In On Medicare Drug Negotiations. https://www.kff.org/health-costs/poll-finding/public-weighs-in-on-medicare-drug-negotiations (12 October).

7. Sachs, R. (2021). Understanding the new drug price reform deal. *Health Affairs Blog*. https://www.healthaffairs.org/do/10.1377/hblog.20211104.184553 (4 November).

8. Brennan, Z. (2021). CBO warns of almost 60 fewer new drugs over three decades with Medicare drug negotiations bill. *EndPoint News* (26 August).

9. Philipson, T.J.; Durie, T. (2021). *Issue Brief: The Evidence Base on the Impact of Price Controls on Medical Innovation*, University of Chicago, (14 September).

10. Congressional Budget Office (2021). Research and Development in the Pharmaceutical Industry. https://www.cbo.gov/publications/57126#_idTextAnchor000 (1 April).

11. Mulcahy, A.W., Whaley, C.M., Gizaw, M. et al. (2021). International prescription drug price comparisons. Rand Corporation. https://www.rand.org/pubs/research_reports/RR2956.html (21 January).

12. Bermonte, A. (2021). Pfizer CEO blasts America's drug pricing system: We have a problem here. *Yahoo Finance*. https://finance.yahoo.com/news/pfizer-ceo-blasts-americas-drug-pricing-system-145734621.html (28 October).

13. Dean, H. (2015). Impose price controls on drugs? *New York Times* (18 September).

FINAL THOUGHTS

THE RESEARCH done at the National Institutes of Health, universities, and research institutions is invaluable. This is where a good deal of the basic science needed to start the R&D process is carried out. Essentially, they generate hypotheses about the root causes of diseases or the potential of a therapeutic approach. But these are just hypotheses. It is the applied research done in the biopharmaceutical industry that validates or disproves these ideas. Only the private sector has the resources to carry out such difficult work. Nowhere else is this done.

This work is extremely difficult. Fewer than 1 in 10 projects taken on by the industry actually lead to a new medicine. And it is not a job that provides instant gratification, as the entire process usually takes 8–12 years. Finally, it is extremely expensive research. Analyses estimate that its total R&D costs for the delivery of a new drug from start to finish is approaching $3 billion [1]. But these factors should only play a small role in the pricing of drugs. Price should be driven by the *value* that the drug provides to patients and the healthcare system.

Regardless of how much money and effort a company spends to develop a new drug, if that drug extends someone's life a mere 90 days, is it really worth $10000/mo? Just because a drug is novel does not mean that it is entitled to premium pricing, especially if it is really no better than existing therapy. Regeneron and partner, Sanofi, learned this when they launched the colorectal cancer drug, Zaltrap®, at a price double that of the already marketed Avastin® (Genentech). When oncologists at Sloan-Kettering Memorial Cancer Center said that they would not prescribe Zaltrap® due to its high price, Regeneron/Sanofi promptly lowered their price to more closely match their rival [2]. Value should be rewarded, but only when justified.

Of course, a single pill of Sovaldi® does not cost $1000 to manufacture. But, even at that cost, it is a bargain when one considers it cures hepatitis C – one of the leading causes of liver cancer. An eight-week treatment of hepatitis C with AbbVie's Mavyret® costs $26400 (Chapter 1). Contrast with the cost of treating liver cancer, which costs hundreds of thousands of dollars despite the fact that cures are rare for this type of cancer. This and other drugs add value.

Pharma and Profits: Balancing Innovation, Medicine and Drug Prices, First Edition. John L. LaMattina.
© 2022 John Wiley & Sons, Inc. Published 2022 by John Wiley & Sons, Inc.

A gene therapy priced at $2.1 million at first sounds outrageous (Chapter 7). But when that therapy cures the acute lymphoblastic leukemia of a two-year old thereby restoring that child's life, incalculable value is generated.

When the world was faced with the most challenging healthcare crisis in a century, it was the biopharmaceutical industry that delivered in breathtaking speed. The industry brought us not just multiple vaccines but also therapeutics like antibodies and antiviral drugs. The advent of Pfizer's game changing Paxlovid™, which reduces the risk of death in those who have already contracted Covid-19 by almost 90%, will contribute to getting us beyond the pandemic far quicker than we could ever have imagined in March of 2020 [3]. It was a healthy and vibrant biopharmaceutical industry that made this happen.

Also, we must not forget that drugs have built-in cost controls – patent expirations. Unlike surgical procedures and other healthcare costs that invariably rise annually, drug prices are guaranteed to drop after their patent life. That adds tremendous value to life-saving medications.

Unfortunately, measures are being considered that will diminish the capabilities of what is perhaps America's most important industry. Make no mistake about it. Impacting the revenues of a drug company will have a direct and measurable negative impact on the R&D activities. Less income = less R&D. How ironic is it that at the very time when the industry is demonstrating its importance, steps are being taken to fetter its capability.

The importance of the biopharmaceutical industry has never been more apparent. Let us embrace and nourish it, rather than injure it by limiting its revenues.

REFERENCES

1. DiMasi, J.A.; Grabowski, H.G.; Hansen, R.W. (2016) Innovation in the pharmaceutical industry: new estimates of R&D costs, *Journal of Health* Economics; 47:20–33. doi:https://doi.org/10.1016/j.jhealco.2016.01.012.
2. Pollack, A. (2012). Sanofi Halves price of cancer drug after Sloan-Kettering rejection. *New York Times* (8 November).
3. Pfizer Press Release (2021). Pfizer's Novel Covid-19 Oral Antiviral Treatment Candidate Reduced Risk of Hospitalization or Death by 89% in Interim Analysis of Phase 2/3 Epic-HR Study, (5 November).

INDEX

Pharma and Profits: Balancing Innovation, Medicine, and Drug Prices, First Edition. John L. LaMattina.
© 2022 John Wiley & Sons, Inc. Published 2022 by John Wiley & Sons, Inc.